D1118598

RECLAIMED POWERS

RECLAIMED POWERS

Toward a
New Psychology
of Men and Women
in Later Life

DAVID GUTMANN

BASIC BOOKS, INC., PUBLISHERS • NEW YORK

Library of Congress Cataloging-in-Publication Data

Gutmann, David, 1925–
  Reclaimed powers.

  Bibliography: p. 317.
  Includes index.
    1. Aged—Psychology. 2. Aged—Cross-cultural
studies. 3. Parenthood. I. Title.
HQ1061.G87 1987   305.2'6   87–47516
ISBN 0–465–06864–2

Copyright © 1987 by Basic Books, Inc.
Printed in the United States of America
Designed by Vincent Torre
87 88 89 90 RRD 9 8 7 6 5 4 3 2 1

TO JOANNA
who deserves much more

CONTENTS

PREFACE

Our times are marked by a special sense of mission, aimed at the redemption of hitherto stigmatized groups. Perhaps as a reaction to the Holocaust of the Jews, which showed us to what obscene lengths a popular prejudice could be carried, our various media and activists seek out burdened peoples—whether women, blacks, homosexuals, schizophrenics—and reverse their sign from negative to positive, endowing them with special grace to compensate for their prior degradation. For better or worse, the aged have entered this special third world of the redeemed and have, as an inevitable consequence, become the subject of ideologies as well as of medicine, social work, and science. Inevitably, the aged have generated their own political activists, and gerontology—the study of the aging process—has likewise become covertly politicized. As with other fields devoted to the study of the stigmatized, ideology begins to determine the questions, the methods, and even the findings of gerontological research, particularly in the social sciences. The current emphasis is on the special victimization of the aged at the hands of an ungiving society. The existential and irreversible burdens of later life are interpreted politically: The inescapable miseries of terminating flesh are externalized to become the sins of capitalism in its last stages. The stigma is transferred from the condition of aging to its social environment.

The field of aging studies had its own infancy back in the 1950s, when I first entered it. I did not choose gerontology to assuage some private guilt about persecuted elders but to build a career and to satisfy my curiosity about a large sector of the life cycle, an empty space on the map of developmental and dynamic psychology. This text, the product of my later years, is still informed by the same youthful ambition: to fill that empty space with significant, detailed features. In the service of that unrelinquished ambition, I have sometimes sinned against the strict canons of science. This text is crammed with points where, without apology, I forage, interpret, and generalize at times far beyond the scope of my data. The decencies of science are indeed violated; but sometimes—like a tank commander who, heedless of flanks and supplies, plunges toward a crucial objective—I reach an insight that I would not have achieved by more careful means, one that is worthy of the most careful testing with all the instruments of quantitative science. One such insight is the theory of the parental imperative and its relationship to later-life psychosocial development, an idea presented at length here.

We study what we are afraid of; and gerontologists are no exception. So mine are the sins of overambition, the usual weakness of unrelinquished youth, but I have not, to my knowledge, committed the modern errors: I have not trimmed my language, methods, or findings to the winds from any political quarter. I have tried to follow the data wherever they have led; and I have not, for the sake of fashion, depicted my subjects as empty, unchoosing victims of social forces over which they have no control. I have tried to present elders as people who have appetites, who make wise and unwise decisions that reflect their appetites, and who bear some responsibility for even the most disastrous outcomes.

In this spirit, and in this book, the later years will not be considered in isolation. Our elders are not an indictment of our society, and the terminal years are not a mere gratuity, an extra budget of days or sorrows. These years will not be viewed as an accidental product of affluence or of improved medical practice, but as an integral part of the total human life cycle, a period that is rooted in and extends individual, social, and species history. As such, the second half of life will be viewed neither catastrophically nor sentimentally; it is neither an exclusive tale of losses nor a golden age. Like any other sector of the life cycle, it has its own distinctive depletions, it has its own gains, and it makes its own special contribution to individual development, social cohesion, and species continuity.

This book marks a station in a long journey, an investigation into the psychology of aging that began at the outset of my graduate career, and that has continued until the present, when my own aging becomes a prime object of my study. I have investigated these matters in American cities, in psychological clinics and wards, in remote villages, in deserts and sheep camps. The empty space gets filled in, but each accomplished map suggests a new terrain—the original curiosity finds a new purchase. At various points others have showed me new roads to travel or have joined their curiosity to mine. I was still a boy in the professional sense when my father-in-law, the anthropologist Robert Redfield, showed me what the awakened mind can do with the dangerous excitement of fieldwork. In my own way, I have tried to give psychological content to the same matters, having to do with social change along the folk-urban continuum that he first explored. Bernice Neugarten trusted me, untrained, with her data, and got me started on a line of research that later became my lifework. As she redlined my first incoherent research reports Bernice Neugarten became—and continues to be—my internal editor. By the same token, Jerome Grunes, M.D., a wise student of psychoanalysis and aging, countered my urge toward overgeneralization. He reminded me of the obdurate uniqueness of each patient and of the need to meet all patients in terms of their uniqueness. He has become my internal supervisor.

I take this opportunity to acknowledge the students who taught me how to be their teacher and who joined me, at some risk, on various field trips. Dr. Jeffrey Urist, Dr. Gary Clevidence, Jennifer Clevidence, Dr. Wendall Wilson, and Dr. Sally Haimo worked with me among the Navajo. Dr. Yossi Ben-Dak of Haifa University first introduced me to the Druze, while Miriam Sonn and Dr. Adina Kleiman joined us later, for the investigation of Druze women.

I must also acknowledge the indigenous scholars, interpreters, and guides who directed our uncertain steps among their people. Dr. Alfonso Villa-Rojas took me to the Maya, and the Mech family of Pustunich (Yucatan) were generous with housing, native food, and information. Max Hanley of Tuba City was our chief informant and interpreter among the Navajo; Howard McKinley, Jr., translated with strict precision from Navajo to English; and Charlie (Choo-Choo) Bracker, champion bull rider, taught me how to track old Navajo into the remotest reaches of their reservation. Harry Bilagodi, Isobel Onesalt, and Florence Esplain also provided good guidance in difficult country.

Kassem Yussuf Kassem was, and still is, guide, sponsor, and colleague among the Druze. I take this opportunity to thank the Kassem family of Rama, Western Galilee, for the many times that I was made welcome in their gracious home. Recruited from a branch of the same family, Adib Kassem and his sister Katif were cheerful and valuable assistants to their cousin and me.

Graduate assistants at the University of Michigan and at Northwestern University helped me organize and report the field data. Dr. Alan Crohn developed a novel approach to the analysis of Navajo dreams; Dr. Thea Goldstine opened up the Navajo Thematic Apperception Test data; and Dr. Urist did the same with Navajo early memories. Sponsored by a generous training grant (#80–14) from the Retirement Research Foundation to Northwestern University's Older Adult Program, Dr. Brian Griffin, with the help of Leslie Groves, brought out unsuspected but crucial regularities in Druze protocols, findings that form the basis of a significant chapter in this text.

It is common practice in acknowledgments of this sort to insist that the typist rendered invaluable assistance and was only slightly less important than the author in getting the book out. As regards Bonnie Lovejoy's contribution, such a statement would be entirely correct. I might have completed this book without her help and dedicated nagging, but I am very glad that this dread possibility never came to pass.

The spouse is vital to all phases of any study; more so when field work is a central part of the project. In my case, Joanna Redfield Gutmann proved to be a true inheritor of her family's tradition in anthropology. She used unfailing ingenuity to create, under varied and trying conditions, centers of warmth, nurture, and even education for our children and me. Stephanie and Ethan Gutmann have been able, in their maturity, to tell us how much our field trips troubled their early lives; but they have also told us how much was salvaged by their mother's resourcefulness and wit. This book is dedicated, with love, gratitude, and respect, to Joanna.

—DAVID GUTMANN
Chicago, Illinois
June 1987

RECLAIMED POWERS

1

ELDERHOOD: IN THE SPECIES, SOCIETY, AND PERSON

Confound not the distinctions of thy life which
nature has divided; that is, youth, adolescence,
manhood, and old age; nor in these divided periods,
wherein thou are in manner four, conceive thyself
but one. Let each division be happy in its proper
virtues, nor one vice run through all. Let each
distinction have its salutary transition, and critically
deliver thee from the imperfections of the former; so
ordering the whole, that prudence and virtue may
have the largest selection.
—SIR THOMAS BROWNE, 1605–1682

There are universal practices and institutions that mark our species
as unique. We invent language, we invent culture, we invent religion,
we maintain an incest taboo, and we are one of the few animal groups
in which an active population of aged individuals lives long after their
reproductive prime has passed. Is this human longevity a matter of
accident—lucky survivors stretching out their years—or is it a matter
of evolutionary design? Do the aged exist among us as a side effect
of our advanced technology in medicine and agriculture, or do they
represent a vital subplot in the larger human story?

3

Human Aging and Human Evolution

While there is no doubt that effective medicine and adequate food supply can help to keep elders alive, human longevity is clearly based on other conditions that are not defined by the achievements of technological civilization. Numerous ethnographers have noted the extreme longevity of isolated mountaineers who inhabit rugged terrain far from clinics, reliable agriculture, and assured food supply. And while this longevity goes with rugged conditions it is not confined to a unique people or to a special region: Pockets of reputed centenarians are found in the mountain walls of Ecuador, the Caucasus, and the Himalayas.

While some sociologists have told me that elders did not exist as a social phenomenon until they chose to study them, the fact remains that the aged were a recognized part of human existence long before social scientists or nutritionists came into being. In greater or lesser numbers, depending on their social and physical circumstances, we find the aged acknowledged as a regular part of the social scene in almost every human community from the most primitive to the most advanced. The normal span of life, according to that ancient text, the Bible, is three score and ten. Much cross-cultural and cross-historical evidence supports a major argument of this book: We do not have elders because we have a human gift and modern capacity for keeping the weak alive; instead, we are human because we have elders.

Such a statement may appear strange, particularly to American readers, who think of the aged mainly in catastrophic terms: abandoned by their families, victimized by society, their bodies depleted. The fact is that there is a strong face of aging, and the durability of the aged is not the result of their stubborn refusal to die, but rather because elders are necessary to the well-being of all age groups, particularly the young. They fill unique roles, vital to the continuity of their extended families and larger communities, across the range of human societies. To repeat, our elders do not survive by chance, nor because of the ingenuity and generosity of the young; they seem to exist by necessity and by design, rather than by luck or special benevolence. The same evolutionary design that has made the presence of elders a permanent feature of our species is recapitulated in individual lives, as aging men and women develop new, hitherto-unclaimed capacities that enlarge their lives and that fit them for their special social and species assignments.

Despite all the attention currently paid by social scientists to the aging process, the developmental possibilities of later life are rarely taken into account. In fact, for the most part, they are determinedly overlooked. The conventional psychology of aging is almost completely devoted to a study of its discontents: aging as depletion, aging as catastrophe, aging as mortality.[1] The typical psychology text on life-span development begins with a chapter about infancy and ends with a chapter titled "Aging and Death." According to this gerophobic view, the aged do not have special gifts, unique to their cohort; instead, they are like blind men who compensate for the loss of sight by increasing the acuity of other, still intact senses. In other words, at best the aged are deemed barely capable of staving off disaster; but they are certainly not deemed capable of developing new capacities or of seeking out new challenges, by their own choice (and even for the sheer hell of it). Whereas in more traditional societies the aged are selected—sometimes self-elected—to face profound spiritual dangers, we North Americans assume instead that the aged are barely capable of warding off the unprovoked blows of fate and that they will usually require help from social agencies to shore up their failing resources. Within this accepted view, psychological development, the emergence of new executive capacities according to a predetermined schedule, simply does not take place in the third age.

By contrast, the intent of this work is to make more manifest a particular evolutionary design of our species and the developmental character of aging: the design that reliably delivers up, under facilitating social conditions, the replacement population of vital and useful elders. In short, I mean to put forward the developmental position as a necessary, remedial alternative to the depletion perspective—the *weak* face of aging—that currently dominates conventional thinking in gerontology.

The Human Elder and Human Parenthood

The catastrophic or gerophobic view is largely based on studies carried out in Western and secular societies, where the aged do tend to be socially disadvantaged and psychologically distressed. But those of us

5

who have studied aging in more traditional folk societies, the natural gerontocracies, see a less gloomy picture of our human fate in the later years. In settings in which elders have the social leverage to arrange matters according to their own priorities, we find striking evidence of new development in both the male and female personalities in later life, the emergence of new executive capacities that go beyond mere adjustment to imposed loss.

Such late-blooming talents are brought to fruition by a maturing process that manifests itself in the later years of individual lives. Since true development must necessarily have a biogenetic basis, we can speculate about the evolutionary engines that could quicken new growth in the later years and about the agencies—species, social, and individual—that are served by these advances.

We know that a minimum amount of at least adequate parenting is a precondition for human development in the early years. Extending this idea, I contend that the state of parenthood, of being parental, is a condition of equal force and dignity, powerful enough to drive the developmental transitions of adulthood, and even those of the later, postreproductive years.

The psychological development of the human elder is founded in a paradox: It is responsive and reciprocal to the unique vulnerability of the child, and to the quite remarkable length of human childhood. The unique longevity of the older human is matched, at the other end of the life cycle, by another uniquely human feature: the long period of early vulnerability and dependency that we term childhood. Most species expand adulthood at the expense of childhood and senescence, but human beings expand childhood and senescence, the pre- and postreproductive stanzas, at the expense of the procreative period, adulthood.

The tendency of *Homo sapiens* to favor the ends of the life cycle over the middle is not accidental, and much of what distinguishes us, much of what makes us distinctively human, flows from this special, tripartite division of the life span. In effect, the long period of childhood dependency and vulnerability depends on a paradox: It requires (and guarantees) a constant brigade of postreproductive albeit still concerned, still responsible elders. Briefly put, the long human childhood comes about because our species has given primacy to new learning over old learning—"automatic" instinctual knowledge. Compared to "lower" animals, our children are born stupid, but they have great promise. Though old learning still manages the vegetative functions of the neonate's internal environment—an infant does not

6

have to learn how to regulate its own heartbeat, or body temperature—at the outset, the nursling is almost completely unable to manage its external physical and social environment. Eventual mastery of those domains requires new learning, painfully acquired while the parents, compensating for the child's inadequacies, manage external reality in its favor. Once matured, the child may acquire and even create knowledge undreamed of by the parents, but until the child's new learning comes on line, normal parents must shoulder a chronic and largely unrecompensed burden. The great measure of freedom that our species has won from bondage to old learning has been gained at the price of adult bondage to parental tasks.

But this chronic parental burden, which I call the parental emergency, cannot be adequately met by the biological parents alone. As modern social history shows, the unsupported nuclear family, no matter how affluent, if it lacks an extended network of kinsmen, servants, and elders, is too often unequal to the task of raising emotionally healthy children. The unbuffered nuclear family does appear to be increasingly incapable of raising children who can avoid addictions, who do not need cults or charismatic totalitarian leaders, who can grow up to be parental in their own right. In order to usher the child successfully through the long period of helplessness, human parents also need to be nurtured. In need of special kinds of parenting themselves, these parents receive uniquely important support (beyond mere babysitting) from the older men and women of their communities. As we will see, the elders' historic assignment has been to maintain the institutional and cultural frameworks of effective parenting. By giving this support to human parents, elders help to protect the tremendous evolutionary advantage of the human cortex, the locus of new learning. Like our weak children, our physically fragile but otherwise potent elders are—again, the paradox—the wardens of our precious human heritage: the potential and the appetite for new learning. The remarkable success of our species is based on the neocortex and on our special parenting practices, which lead to the successful transformation of cortical possibilities into executive capacities of the human ego. And this human parental enterprise is backed not only by biological parents but by the "extended" parents: the parents of the parents, the patriarchs and matriarchs of the human assemblage. In later chapters, I will describe the special developmental axes along which older men and women mature into their special assignments as emeritus parents.

The Conventional View:
Aging as Catastrophe

At this point the reader might be asking, "By what right do you so confidently impose a developmental outline on the aging process, when so many careful and esteemed investigators have not found it, discovering instead ample and convincing evidence for the catastrophic position?" My answer is this: While psychological development takes place in all seasons of the human life cycle, including the third age, we do not observe it there because we have in some cases not found and in others rejected the special theoretical and methodological "lenses" that reveal the subtle tracings of development in the later years. In infancy and childhood, growth takes place in all dimensions, physical, mental, and emotional; it is apparent to all observers, regardless of their theoretical stance. Because early growth is so rapid and unequivocal, even the unsophisticated observer requires no special lenses, no special forms of observation to see it clearly. Insofar as development brings about gross changes in their bodies and basic skills, children themselves are quite aware of their advancement in these spheres. But later-life development does not show itself so vividly; instead, it may cause no more than the quiet ripening of selected mental and spiritual capacities, or a gradual shift in appetites, interests, and occupations. In addition, the subtle signs of later-life development are easily obscured by visible, tangible changes in the aging body, which seems to register decay rather than growth and maturation. Small wonder then that our current psychology of the obvious, with all its emphasis on precise measurement of the gross, observable object, picks up the performance decrements and ailments of the aging soma and misses the less physical, more inferential maturations of the aging psyche. When psychologists look at elders through their materialist lenses, they can see only the obvious losses of beauty, health, intellectual sharpness, and social opportunity. Predictably, they render a model of aging as catastrophe, aging as wasteland.

This patronizing view of the elders as damaged, as incapable of new growth, may be propagated today by the profession of gerontology, but its ultimate sources lie in our history and culture. Gerophobia runs deep in the American grain, part of the founding myth of a nation

8

whose Founding Fathers were in truth Founding Sons, rebellious sons, refugees from patriarchal gerontocracy. Most Americans are immigrants or the children of immigrants; we descend from those who fled the old ways of the "old country," the stifling traditions monitored by forbidding circles of chiding, gossiping elders in claustrophobic villages. America has been a collecting point for this planetary elite of prometheans—self-selected Americans, contra-patriarchal individualists before they ever saw this land—whose private myth has it that they fled the elders of the old land, and in so doing stole their power and brought it with them to fuel their achievements in the new. The immigrant's personal myth of rebellion matched the founding American myth of a revolutionary brotherhood that had broken from the tyranny of a king, the ultimate gerontocrat, and that had shared out his royal mana, his charge of sacred power, among the democratic comrades, in order to create a new, participatory aristocracy. In the United States, the power of the old fathers passed to the society of the young democratic brothers. Individually and collectively, we are the inheritors and wardens—but also the potential victims—of that stolen power, and we have split off our fear of that potency into the aged now resident among us. Thus our covert boast is that the aged will never take back from us the power that they once exercised in the "old country"; our corresponding fear is that they will somehow wrestle it away and turn it against us. Accordingly, we are not comfortable with the aged unless we deny their potencies, seeing them instead as impotent and—like the toppled king—shorn of their mana.

Geropsychologists, products of the same culture and the same myths, enact their special version of gerophobia by elaborating the doctrine of catastrophic aging, by portraying the aged as needy of services, and by playing down the developmental possibilities of later life. In our traditional psychology about aging, we reenact the degradation of the elders that our rebel ancestors first accomplished.

But if the psychic health of the younger American depends to some degree on denying the developmental possibilities of later life, we also pay a high social price for that arrangement, a price that is only now becoming evident as the beneficial influence of strong elders increasingly fades out of our collective life. A major aim of this book is to confirm the developmental dimension that scientific bias, scientific error, and cultural myth have held back from our models of the aging process.

The Triaxial View
of Human Development

Any chapter in development marks the point at which the individual takes possession of some piece of our common evolutionary heritage, expressing it in ways that bear the stamp of local culture, as modified and interpreted by personal character. Each factor that contributes to a developmental advance—whether species, social or personal—is studied separately, through special instruments, lenses which are selectively sensitive to these phenomena.

But before aiming these instruments, I should first specify their particular sensitivities and the natural phenomena that each one brings into focus. The triaxial nature of human development is neatly expressed in this powerful insight from the noted anthropologist Ralph Linton (with Abraham Kardiner, 1945): "In some ways each man is like all men; in some ways each man is like some other men; and in some ways each man is like no other men." With this clarifying statement, Linton provides an escape from academic imperialism, from the need to believe that some anointed viewpoint, whether psychological, sociological, or phenomenological, "owns" the independent variables of human behavior. More to the point, Linton introduces us to fundamental aspects of human existence that are far-reaching in their influence, though perhaps not immediately evident to our senses. In so doing, he points to special domains wherein the clues to later-life development might be found; he helps us to grind the lenses that will aid in the search; and he suggests new criteria for judging the value of new findings.

Linton's levels relate to the thesis of this book in that they refer to distinct domains, to the ways by which distinct human forms are maintained and to the varieties of individual competence and mastery that develop within these frameworks.

Thus Linton's first level ("each man is like all men") refers to the human species as a distinct entity and to the generic human characteristics that contribute to its survival. Specifically, the statement refers to the human body, to the reservoir of elemental appetites and developmental possibilities that underwrite the physical survival of the individual and, through sexual reproduction, of the species as a whole. In addition, Linton's first level concerns the interface at which psyche and soma meet, where the requirements of our bodies for nourishment, for comfort, for pleasurable sensation, and for propagation express themselves in fairly standard ways through the psychological system. The first level refers to the familiar hungers, excitements, and fears that define our human nature, and that are shared by all normally endowed men and women across the human race.

Moreover, the first level refers to a fixed reality of the human situation: to conditions of hardship and scarcity, of real or potential danger and emergency. Ultimately, we are all brothers and sisters under the skin, and this common human nature stands forth very clearly under conditions that threaten personal or species survival. The extermination camp is a poor place to observe the learned effects of social breeding or individual history. But while some dangers bring out the brute beneath the civilized human exterior, other forms of emergency can bring out, with equal predictability, the very best in us. Adult humans, regardless of their social and ethnic origins, can become strikingly self-sacrificial in wartime or in the service of their children. Indeed, an essential argument of this book is that men and women are both (and possibly equally) dominated by the state of parental emergency, and that important psychological changes in the later years are driven by the same forces, the same kinds of species imperatives that bring about the gender distinctions and parental sacrifices of the earlier years.

In sum, Linton's first level finds its best metaphor in geophysics, in the continental plates that shift and grind in their slow march deep below the earth's surface. They have never been directly observed, and their very existence can only be inferred from surface phenomena. Nevertheless, in their tectonic action, these vast, crawling plates supply the basic energies that sculpt the major features—ocean basins, rift valleys, volcanoes, mountain ranges—of our visible, ever-changing landscape.

Linton's second level ("each man is like some other men") refers to the collective and cultural frameworks of human existence. A society includes those who either share common understandings or, at the very least, share common ways of expressing their understandings, however different these might be. Linton's second level also refers to those trained capacities that individuals devote to preserving the social order and to preserving themselves as social beings. At Linton's first level, we see the individual as an expression of raw, insistent nature, ruled by the aggressive instincts that ensure individual survival and by the sexual instincts that ensure species survival. But at the second level we view men and women as standard representatives of social learning, of the indoctrination that defines and preserves a social group across the generations of individual lives. Seen from this level, the individual no longer appears to be driven by the instinctual imperative, by some innate nature, but by externally imposed requirements stemming from social nurture: the language, attitudes, and customs of a particular human group.

While a complete maturational cycle can take place only within socially organized settings, the lenses fitted to these settings do not pick up developmental phenomena as such. At this level of observation, the species imperatives are easily confounded with (and rephrased into) the idiom of the social imperative; and the effects of human nature seem to be overwhelmed by the influence of social nurture. When we hold imposed social forces responsible for all human thought and behavior, it is akin to holding wind, rainfall, and wave action entirely responsible for the shaping of the earth's surface, while ignoring the effects of earthquakes, volcanic action, and the slow wearing of continental plates. Rain and waves do erode mountains, but they do not raise them. The ranges are lifted by the less visible but nonetheless mighty forces of the earth's interior. Sociologists, the students of second-level phenomena, too often ignore the effects of psychological earthquakes and attribute all shaping power to the effects of our social "weather." Studying individuals in one society, it is quite easy for them to confuse native drive with external social pressures, or to mistake the innate rhythms of intrinsic development for the pace of socially induced change. The "tectonics" of the species level are easily misread as second-level cultural phenomena and are conceded to elements within the social bath: the webs of social opportunities, restrictions, and "support networks" that surround the socialized individual. Such confounding of the species and social influences can easily grow out of studies restricted to single societies, pitched exclusively at Linton's second level.

Linton's third level, concerning the ways in which "each man is like no other men," refers to the unique as opposed to the social self, to the means by which we achieve a sense of specialness within some recognizable social framework and to the ways through which we preserve our sense of uniqueness. It also refers to the part that defining events or agents—inherited temperament, accidents of birth, special endowments or disabilities—can play in altering the trajectory of individual development away from species or social norms.

I use "norm" in the statistical rather than the qualitative or moral sense. Norms are abstractions based on the common characteristics of groups rather than actual persons, and individual variations on the collective motifs are both inevitable and even desirable. The human fingerprint provides the best metaphor for this level: It is a unique feature of our species, immediately recognizable as part of our general nature, yet each of us has a fingerprint so distinct from all others that it can be used as a form of identification, unchallengeable in any court of law. By the same token, the sense of personal identity or selfhood is likewise a kind of fingerprint. As our psychological signature, it is based in the qualities that we share with certain respected others but are also unique to us. At the third level, we study the unique individual as we would a novel or a poem. The lenses crafted to this level pick out the special texture and idiosyncratic origins of a particular life, while blurring and diminishing the larger social, species, and historical contexts from which it also derives.

The Sequence of Development

Before applying Linton's distinctions to our subject matter, and before examining particular stanzas of human development in later life, let us briefly review the characteristics of a typical developmental sequence. There are as many definitions of human development as there are behavioral psychologists, but most of the latter would agree that any successful maturational sequence has as its outcome some new executive capacity or capacities, some hitherto latent means, whether physical, mental, or emotional, for ensuring greater mastery over the problems and challenges posed by the inner life, by other people, by social institutions, or by the particular developmental advance itself. In the psychological domain, a developmental sequence moves through three major subphases, in the course of which the emerging capacity is given sharper focus, refinement, and power.

I term the first stanza the *surgent*, or eruptive, subphase of development. In this initial growth stage, and particularly in the early years, change is explosive. The body, urged by its intrinsic "clocks" or timetables, churns out new potentials, toward new appetites and new learning, and toward the new forms of social participation that these imply. At this early point, the established personality system of the individual, based on familiar, "normalized" appetites and skills,

13

shrinks away from the raw energies and denies these alien promptings free access to action or to public display. Consequently, the new impulses announce themselves mainly in indirect ways, through coded, seemingly incoherent language that escapes the ego's vigilance: dreams, private reveries, and body languages such as slips of the tongue and involuntary gestures. Here the fantasy function plays its assigned, adaptive role in the personality system: By supplying an alternative route for their discharge, fantasy keeps potentially dangerous tendencies out of behavior. The surgent impulses are contained in fantasy, and their potentially dangerous consequences are rehearsed and planned for before they are released in the objective world.[2] In fantasy we can punch out our boss, while still holding onto our job in the real world.

But all psychic contents move, however haltingly, toward direct communication via the common language and conventional behavior. Unless they stimulate neurotic repressions (and so remain locked in the unconscious), the eruptive drives, as they are tested in fantasy, become partly detoxified so that they can be shown in overt behavior. Eventually, they break out of their confinement in fantasy and seek real targets in the objective world. Like flood waters that cut new riverbeds, the surgent energies reach out. They create new psycosocial conditions, they engage the sponsors (the guides, models, co-actors) who will further their transformation, from raw energy into executive capacities of the personality. Thus, as they move toward consciousness and direct expression, the emerging forces take the form of cathexes (Freud's term for the psychic investment in relationships) and attach themselves, depending on the stage of life, to parents, siblings, peers, teachers, mentors, and lovers. The child's teddy bear is the precursor of the adult's lover. At this critical point, the developmental process has moved into the *proactive*, or reciprocal subphase: Other individuals—for example those engaged in the same developmental struggles or those who have already passed through them—are enlisted to validate and to name the raw, still suspect potentials. The emerging, sometimes awesome powers are dangerous excitements. Accordingly, trusted sponsors are required to normalize the surgent potentials, to demonstrate that they will not destroy the self or its partners, and that their sign is positive rather than negative.

In addition, the task of the sponsoring other is to introduce the developing individual to larger, more public domains, those beyond the intimate worlds of child and parent, or pupil and mentor. Thus the sponsor in the proactive phase has the added task of initiating the final stage of the developmental sequence. A teacher, for example, implies schools, and so the teacher points the growing individual, quite naturally, toward the larger, institutional contexts of their personal relationship, thereby ushering in the *sculpted* subphase. In this last act of the developmental journey, the growing individual moves beyond one-to-one relationships and their idiosyncratic rules, to encounter a general consensus: more impersonal, more abstractly social worlds and their objective laws. Now, instead of parents, the individual deals with families; instead of particular friends, with peer groups; and instead of single opponents, with opposing, competing teams. Despite their internal origin, the burgeoning aptitudes are shaped at this point by external requirements, by the rules of the game, as these are communicated by responsible (and responsive) elites. But while the social rules now chisel and shape the native capacities, the growing individual relates selectively to the impinging regulations and guides conforming behavior according to some personal interpretation of them. In effect, the individual uses the rules as much as the rules shape the individual. The child, for example, learns the family's language, and thereby submits to the rules of syntax, but that same child stoops to conquer. By mastering the rules of language, the child brings surgent issues into clear communication and gains some control over the new social domains that are open to the language user. The paradox of development is that new capacities have to be socially articulated in order to be available, as resources, to the developing self. Like the human fingerprint, which is at the same time generic to the species and unique to the individual, the socially sculpted, fully developed capacity becomes simultaneously a resource for society and the signature of personal identity.

Development and the Linton Levels

Human development is essentially Lintonian in character, for as maturation progresses through the surgent, proactive, and sculpted subphases, it at the same time ascends the Linton levels. Thus the surgent phase, like the first level, is fueled by protean energies that are generic to species, to gender, and to life stage. The proactive stage corresponds to Linton's second level, in which the developmental potentials have found social objects and are being poured into their social molds. They begin to acquire the tone of normalcy, as well as distinctive form and direction. The sculpted phase corresponds to Linton's third level. At this point the individual trades grandiose illusions of omnipotentiality for the attainable goal, of personal uniqueness. The maturing individual puts a unique stamp, a fingerprint, on the newly minted executive capacities that have received their dynamism from human nature and their rough hewing and shaping from social nurture.

In short, as a developmental sequence moves through the Lintonian stations, individual maturation recapitulates the broad sweep of human development—from a brute, driven reflexively by peremptory hungers and undifferentiated urgings, to a self-determining person, integrating individual purposes with collective requirements.

This correspondence between the stanzas of a typical development sequence and Linton's three levels suggests a set of standards for judging the adequacy of any maturational process. The fully matured executive capacity will serve (though not to the same degree in all cases), to ensure self, social, and species continuity. Thus seemingly accidental events within the lives of single individuals may be seen to have a "species" significance and a larger meaning. In later chapters, we will see how the normal, surgent changes of later life can be transformed into true developmental achievements, insofar as they meet and satisfy the special requirements placed on the older person by our conjoint idiosyncratic, social, and generic human natures.

The Mind/Body Schism in Contemporary Psychology

While a true developmental outcome combines the three aspects of man, in such a way that each can be read in the others, contemporary geropsychology, and the social sciences generally, tend to pull them asunder. Mind is separated from body, and the dynamic, purposeful organism disappears from our theoretical radar screens. The conception outlined here, of the triaxial life cycle, is at odds with the prevailing view in Western and particularly North American psychology. In general, driven by an individualistic, egalitarian appetite, American psychology opposes the idea of an inflexible human or "species" nature. Rejecting this essentially tragic view of man, our optimistic psychologists concentrate instead on the presumably reversible effects of social nurture. They deny Linton's first level and disparage its spokesmen as front men for reactionary politics. According to the radical and humanistic psychologists, the Freudians and sociobiologists reserve for the first level, for the body and its genetic makeup, the shaping of behavior that properly belongs (according to the radicals) to society, or (according to the humanists) to the unique, self-generating self. The spokesmen for Linton's first level concede to immutable human nature (and to the genes) important qualities, particularly those having to do with gender differences, that activist social scientists insist on seeing as the artifacts of social oppression, to be revoked through revolutionary action.

Conventional geropsychology has faithfully followed the lead of American social-behaviorism. The ultimate dynamism for development has been taken away from the species and conceded almost totally to outside social forces. Accordingly, for the social geropsychologists, cultural and economic dynamics become the primary engines of human development. The aging individual is visualized as an incidental product of a particular economic system, value consensus, and collective history. While denying the first level, established theorists and researchers in geropsychology ascribe great and independent power to the sociologist's variables, such as culture, social class, age-cohort. They externalize the influences that shape the behavior, the attitudes, and the appetites of the older person away from his human nature and into the conditions of social nurture that surround him. The older person is seen as a *tabula rasa* upon which the local culture will write its special text. The body, as a product of a common evolutionary history, and as the dynamic source of generic energies, appetites, and developmental timetables, disappears from the theoretical calculus and is preserved only in the most negative sense: as a collection of defective organs, on the edge of death.

Thus, urged by their intellectual hubris, modern psychologists elevate the mind over the body and depict humankind as ahistorical and self-created, a product of the same social forces that the human mind has itself brought into being. However, by trying to escape the constraints of our mortal flesh and our evolutionary history, we arrive at what has been called the oversocialized view of man in the social sciences, and nowhere is this more evident than in the discipline of geropsychology. The aged have been reduced, by academic fiat, into their social equivalents—the social forces that act on them. By depriving elders of any shared, willed responsibility for their fate, this sociocentric bias confirms the catastrophic picture of aging. As a self-fulfilling prophecy, it adds to the victimization of our elders.

More recently, the flight away from developmental and genetic determinism has led some social scientists and geropsychologists to elevate Linton's third, or idiosyncratic, level of human organization above all others: "Each man is like no other men." Celebrating the unique, ahistorical self, these romantics invoke the aleatoric principle of development, asserting that life is a random walk and that there are no generic standards against which individual progress can be judged. Late-adult development becomes relativized and nonnormative, and each free adult presumably determines the criteria for judging his or her own growth. In effect, by ruling out the possibility of general, species norms for development, the humanistic psychologists have democratized the business of defining standards and have conceded to each individual the right of self-determination in regard to the values and standards by which he or she will be measured. Consensus, mores, and social norms are undone; each life is a novel, and there is nothing left for psychologists to study but individual life careers, in all their uniqueness and in terms defined not by the investigator but by the subjects themselves.

At first glance, this appears to be a liberating perspective. The humanistic phenomenologists have seemingly freed us from the coercion of instinctual, social, and historical imperatives. But as this most murderous of centuries should teach us, well-meaning liberation movements can bring about new and worse forms of servitude. Humanistic psychology may seem to free us from oppressive determinism, but it also undercuts the standardizing, ordering work of science and leaves us all, the aged included, vulnerable to charlatans. For when clear theoretical criteria are lacking, it is difficult to sort out true advances from other forms of later-life change or to evaluate the often conflicting claims of self-elected, self-promoting "change agents." Thus we have reached the point where almost any variety of nonpsychotic psychological change in later life can be glibly termed "developmental." Inevitably, the lack of normative and theory-based criteria leads to a lack of professional standards. We have entered the age of self-ordained healers. The task of sponsoring so-called "growth" is taken over by all the Dr. Feelgoods, from sex therapists to affluent mystics, who claim that they can make older people feel better. As in many fields of psychology, the study of intricate, paradoxical processes of development and adaptation is shunted into the study of what is now called self-actualization or instant self-enhancement. The new treatments proliferate: self-help groups, reminiscence therapy groups, bereavement groups—all promising unspecified "growth" and self-fulfillment. Although true growth, like any birth, must entail risk and pain, the tragic sense of life is denied, and the study of development is confounded with the study of "whatever works" or "whatever feels good" to the isolated individual. Consequently, the "life satisfaction" scale becomes the ultimate measure of developmental achievement in later adult life.

However, the politics of narcissism do not travel well into later life. When we judge the season of elderhood from this hedonic bias, it does not come off well, no matter how much we tinker with scales and criteria. When the new psychologists, the apostles of liberation, poultice the narcissistic wounds of their usually younger constituencies, they only add to the narcissistic wounds of our elders. Their "humanism" ends by making the older person particularly vulnerable to the general, cultural phobia against aging. When spokesmen for the unique and imperial self consider aging from their favored perspective, of idiosyncratic individualism, the second half of life appears as a wasteland. It contains few obvious possibilities for sharp pleasure, for new experience, for unfettered "growth." Instead, it stands out as a time of loss and retrenchment, and the catastrophic view is again confirmed.

Toward a Triaxial Conception
of Later-life Development

If growth does indeed take place in the second half of life, Dr. Feelgood cannot help us recognize it. We cannot pick up the clear, unequivocal tracings of developmental tectonics at Linton's third level, for there the principles of chance and idiosyncratic choice are elevated over all other forms of determination. Nor can we find it at Linton's second level, where the forms of social nurture are given first priority, and the powers of universal nature are obscured and distorted by the effects of local custom and usage. In order to comprehend development at any age, and particularly in later life, we have to employ the discarded "species" lens and focus it on phenomena that gerontologists, with few exceptions, have thus far disregarded.

This book takes up the unfinished business of developmental geropsychology. Other investigators have fixed, almost exclusively, on the social and idiosyncratic levels of personal organization in later life. But the first, or species, level, whence comes the basic energy and essential direction of any developmental advance, has never been brought under the necessary scrutiny of comparative, transcultural investigation. Nevertheless, since development is rooted in our generic nature and serves social and species goals as well as personal ones, studies conducted at this basic level could highlight the special evolutionary assignment of the human elder.

Such investigations are risky. Linton's first level is based on surgent though covert processes. To study it, we need to venture out of our comfortable social milieu, to confront both the intrapsychic and extracultural realms of experience. We need to study the hidden tectonics of human personality as these are manifested across the various continents of the human condition.

More precisely, we are required to study the motives and appetites of older men and women, particularly when these are in their "pure," that is, still unconscious or surgent, state. We usually study the unconscious drives through the medium of the projective tests, which elicit their fantasy derivatives. These instruments, described in detail in later chapters, consist of relatively unfamiliar stimuli: inkblots or ambiguous pictures. As such, they do not provide the subject with much anchoring for a conventional, socially appropriate interpretation; accordingly, the more unsocialized, unconscious themes are likely to emerge in the subject's interpretations of these pictures.

Returning to our geophysical metaphor, the projective tests are like psychological seismographs. They pick up signals from the deep strata of the psyche, those that grind away down below the masks of public behavior and conscious awareness. Working across a range of disparate societies, varied in their cultural contents, I have used such instruments to pick up the rumblings of unconscious processes, the prediction being that the tracings elicited by these instruments would distinguish older from younger subjects in characteristic ways, regardless of the cultural gulfs between them. Clearly, any psychological tendencies that varied mainly by age (or by age and gender) across our spectrum of societies must have to do with life stage (which has been held constant) rather than culture (which has been systematically varied). In other words, we glimpse some transhuman order, the forest instead of the trees, only when we look beyond culture as such. It is then that the larger regularities announce themselves, in the form of predictable sequences, across a wide range of racial, social, and physical settings. In effect, both the species and the surgent levels are best revealed through equivalent strategies, by reducing social constraints or by randomizing them, so that no set of social rules will throw our data out of balance. Through the use of "seismographic" projective instruments, used comparatively, we sift out the otherwise undetectable but powerful signals from the underground, species domain. We may thereby identify surgent energies, forces that initiate the growth sequences of later life and that underwrite the elder's role in human parenthood.

The Development of the Argument

This book will draw on two separate pools of cross-cultural data: my own comparative studies of younger and older men across a variety of preliterate agricultural societies, and a body of ethnographic accounts that cover a wide spectrum of cultures. Though written by independent investigators, all these accounts devote some attention to the roles and behaviors of older men and women. Nevertheless, in contrast to my own studies of aging, these anthropological data were not collected according to a comparative design, nor were they, for the most part, collected by ethnographers with a primary interest in aging. Thus, while anthropologists have routinely approached the aged as their best sources, they have typically used them as informants about their own cultures. In fact, ethnographers have only recently discovered the aged as a special population, worthy of study in their own right. Thus, for most anthropological citations in this work, the references to the aged have been lifted out of their larger settings in society and the texts themselves. The use of such randomly acquired, anecdotal material is questionable; therefore, the ethnographic data from other investigators will not be used as primary source material but will be used illustratively, mainly to give a sense of the ways in which universal themes might be expressed under various social conditions.

The main text has three major divisions, which correspond

roughly to the order of my studies and to the evolution of my ideas. I begin with initial hypotheses, brief descriptions of the study sites (as well as the reasons for their selection), the field methods, the nature of the data, and the data analysis. Following that I discuss the major findings from the field studies, with special emphasis on the results gleaned from the analysis of projective and interview materials. Again, the findings from my own studies will be illustrated and extended through citations from secondary, anthropological sources. These ethnographies will illustrate the different manifestations, in various societies, of the developmental trends identified through my cross-cultural comparisons. The remaining text will be devoted to models of the developmental process itself—the ways in which nature and nurture may act together to bring about personally, socially, and humanly desirable changes in the older person.

I first set out some universal aspects of aging psychology, those that belong to and are studied in the species setting; then I move to an exposition of the various ways in which these generic possibilities are crafted, to take up their final individual forms within the particular social context. Human nature supplies the raw wood, but the ordering mind—the mind that comes into its own within society—acts as the carpenter to shape the wood into a thousand forms, as called for by ecological pressures and cultural aesthetics.

The central idea of this book is summarized in chapter 8, "Aging and the Parental Imperative." Though coming late in the text, it is the capstone section, in which I link the species, social, and individual contexts around the requirements and imperatives of human parenthood. At that point I spell out, in detail, the argument first advanced in this chapter—that the aged, as emeritus parents, play a vital part in ensuring the success of this most crucial of all human undertakings.

This is the plan. The reader will judge if the book achieves its purpose of putting forward a hopeful and conceptually useful view of human aging.

2

THE COUNTRIES
OF OLD MEN:
STUDY SITES AND
NATURALISTIC METHODS

So far from the town government affairs are few;
So deep in the hills, men's ways are simple.
Though they have wealth, they do not traffic with it;
Though they reach the age, they do not enter the
   army.
Each family keeps to its village trade;
Gray-headed, they have never left the gates.
Alive, they are the people of Ch'ên Village;
Dead, they become the dust of Ch'ên Village.
Out in the fields old men and young
Gaze gladly, each in the other's face.
In the whole village there are only two clans;
Age after age Chus have married Ch'êns.
Near or distant, they live in one clan;
Young or old, they move as one flock.

—Po-Chü-I, A.D. 811*

The basic ideas leading to the field studies reported in this book
were generated in 1956, in the course of analyzing data developed by
the Kansas City Studies of Adult Life (a research project of the Committee on Human Development of the University of Chicago). Many
kinds of information were gathered from a large community sample
of that most American city, and, as a doctoral student, I was respon-

* From Arthur Waley, trans., *Chinese Poems* (London: Unwin Books, 1946).
Reprinted by permission of the publisher, Unwin Hyman Ltd.

23

sible for the analysis of one small piece of the data pool—the responses generated by a Thematic Apperception Test (TAT) card. The TAT cards were originally designed in the 1940s by Henry A. Murray and his associates at the Harvard Psychological Clinic. The test battery consists of scenes which touch on major life concerns—achievement, intimacy, isolation, aggression, etc.—and the pictures are drawn so as to suggest these matters, while leaving ambiguous the relationships among the various stimulus figures portrayed in the cards. The way the test subjects define these relationships presumably represents their own inner world, their own subjective stance in regard to the card issues. My research dealt with the responses generated by a card designed especially for the Kansas City studies.

Some Guiding Hypotheses

The results of that pilot work (reviewed in chapter 6) led to a hypothesis of developmental change in later life: specifically, that older men and women come to possess psychological capacities that were previously unavailable to them or that had been lived out vicariously through persons of the opposite sex, such as the spouse. The hypothesis held that men acquire, in later life, "softer" qualities of affect and of cognition, which are at odds with the previous hard-edged "masculine" definition of their personalities. In this seemingly developmental advance, the older man seems to gain a sensitivity and tenderness previously lacking in his psychological makeup. Conversely, the inner life of older women, as portrayed in responses to projective tests, appears to move on the opposite tack, as they embrace the aggressive masculinity that the older men are relinquishing. This cross-gender trade-off in psychological contents suggested a natural process, possibly developmental in character. A preliminary review of ethnographic materials supported the Kansas City results, indicating that older women became ascendant across a wide range of societies. The universality of this finding added to the impression of a powerful and possibly developmental phenomenon.

However, further cross-cultural testing was indicated, particularly to rule out the possibility that the observed changes in men, toward mildness and accommodation, represented loss rather than development. While the vigorous older woman is clearly gaining new energy, the mild, pacified aging male could be going downhill; instead of gaining new endowments, the older man may instead be losing those that he once had. If such is the case, then the observed age changes—in men at least—could not reflect development but might be traceable to specific cultural effects, for example, to an American society that degrades the prestige of its older men. In order to control for such parochial influences, it was necessary to test the developmental hypothesis—that the mildness of older men was a specieswide phenomenon, and that it betokened possible growth rather than irreversible loss—in societies wherein older men are likely to gain, rather than lose, prestige and social supports.

Site Selection

A hypothesis concerning human development, particularly one that extends developmental conceptions into the third age, can be adequately tested only across a panel of societies distinct from each other and different from the one in which the tracings of new growth were first noted. By randomizing the sample of societies, we control for possible cultural effects: Given the societal variation, regularities in the aging process that are observed across communities cannot logically be stimulated by the cultures themselves (as these are, in each case, distinctive).

Just as it is required to achieve variety across culture, it is also necessary to ensure constancy and sameness within societies. Accordingly, I chose a panel of communities that, though varied in terms of cultural characteristics, shared a common aspect of social structure: That is, each society maintained a notable degree of intergenerational continuity. By the use of these criteria, I controlled for cohort effects. In the sample of societies, the lifeways of older and younger men alike were guided by much the same traditions and thus were similar in major respects. Young men could see their futures represented in their fathers, and fathers could see their pasts echoed in their sons. (The timeless, redundant quality of the folk-traditional society, in which young and old discover themselves in each other, is captured in the epigraph poem, "Chu Ch'ên Village," written in the ninth cen-

25

tury A.D. by the Chinese poet Po-Chü-I.) Under such conditions of cultural stability, observed differences between younger and older men cannot be ascribed to cohort differences or to social change across the generations; they are more likely to reflect innate change within individuals, over the life span.

Finally, our criteria were designed to control for the effects of particular social supports on psychological dispositions. Because we had theorized that the shift toward more passive, accommodative postures (first observed in American men) had an intrinsic basis and was more than a reaction to social conditions, it was necessary to test this hypothesis about intrinsic human nature in societies wherein older men are likely to gain, rather than lose, prestige and social supports. Accordingly, the study was carried out across a range of "participatory gerontocracies." In these societies, older men gain a measure of prestige or social support or both, not only because of their earlier achievements but also by virtue of their advanced age per se.[1]

In order to give a sense of the peoples that met the selection criteria, I present some distinctive characteristics of the comparison societies in the following section. More detailed accounts, emphasizing both the surgent and ego aspects of the groups' personalities, will be found in the appendix, under the heading "Culture and Personality Among the Navajo, Druze, and Mayan Peoples."

The Navajo

The Navajo reservation, in the southwestern United States, is larger than many eastern states, encompassing parts of northern Arizona, western New Mexico, and southern Utah. Though there are well-watered, well-forested ranges on the eastern side of the reservation, for the most part the Navajo lands are relatively bare of vegetation, punctuated by great, dramatic mesas and suitable mainly for the herding of range livestock—sheep, goats, and horses. The Navajo constitute a thin dusting of people over this high desert, bounded on all major directions by their four sacred mountains. The Eastern Navajo are somewhat more political, Christianized, and Americanized; the Western Navajo, spread along the longitude that bisects the Grand Canyon, are more traditional, more observant of the native lifeways and religious rituals. It was mainly in this western sector, from Navajo Mountain to Tuba City, that we found Navajo populations who met our criteria for gerontocracy and intergenerational homogeneity. The pace of social change has quickened since then, but during the years of our study (1964, 1965, and 1970) there was a continuity of experience between young men and old men; the young man could see in the old man much of what he would become, while the old man could see in the young man much of what he had been. For the Navajos, generational continuity is preserved in the relatively unchanging islands of folk life found mainly to the northeast of the Grand Canyon. There the traditional life of migratory herders is buffered against social change by the deep moats of canyons and wide stretches of arid high plateau.

The Eastern Navajo have a sense of tribe in the abstract, political sense, but the Western Navajo keep the more traditional experience of community: clans and extended families living together in a huddle of wood hogans and corrals, each compound isolated on the bright plains and connected to other such "outfits" by rough trails. In these natural communities of clansmen and kinsmen, the older medicine men, with their specialized knowledge, are the natural leaders. Dignified men, encrusted with turquoise jewelry, these older *hatrahli* (singers) can be imposing figures, and to them are brought for remedy the matters that most occupy the Navajo: personal misfortunes, especially physical illnesses. To the Navajo, such events betoken a basic disruption of the harmony that should exist between individuals and the supernatural influences that surround them. In their view, these ruptures are brought about by individual breaches of taboo, by the action of witches, or by contact with enemy power (the power inherent in aliens, in other tribes or white men).

The Navajo were not always so hypochondriacal. In their earlier history, they fought other tribes rather than wasting diseases. The Navajo are a branch of the warlike Apache who ages ago wandered down from Canadian Athabasca, driving the Anasazi, the more pacific Pueblo Indians, before them, finally penning what are now the Hopi on the unassailable mesas of their present reservation. But today the Navajo have been defeated and pacified by the Anglo, and they can no longer find their enemies and prey "out there," in adversary warriors or their horse herds. Now the people's enemies have figuratively moved inside their bodies, to become destructive illnesses or diseased organs. Additionally, the enemy has been cut loose from earthly figures; if still outside, he has become magical and supernatural—a ghost, a warlock, a werewolf (*yenalchlonje*). Because now even the young Navajo fear an old man's version of the enemy—diseased organs and malignant supernatural forces—the *diné*, the people, seek out powerful old medicine men for remedy, hoping that their chants will help them prevail over bad power. The Hand-Tremblers, who once divined the location of the enemy's horse herds for the Navajo raiders, now pass their shaking hands over the bodies of their patients to locate the enemy within or the witch without, and on the basis of their diagnosis the proper singer is brought in to manage the ceremonial, the specific chant against the indicated disorder.

Through his chants the *hatral* retells the story of a mythic event, of a time when the gods intervened in human affairs with good effect, against troubles similar to those now afflicting the patient. For the preliterate mind, to retell a myth is to reactivate it, to recreate the climate of the gods' interventions, this time in favor of the patient. By chanting their deeds, the singer brings the gods and their beneficent influences to bear on the patient, in order to restore the lost harmony between his client and a piece of offended nature. While a large percentage of the elderly Navajo men claim to control some aspect of the traditional healing arts—a special chant against a particular form of illness or special knowledge of curative herbs—the most powerful medicine men are those who control the widest range of sings, particularly the great Nine Nights chants, such as "Mountain-Top Way," in which the yeibeichei dancers, the simulacra of the gods, appear. Though younger men can be apprentice singers, such knowledge and expertise mainly accrues with age, and so the oldest medicine men tend to have the largest reputations across the Navajo lands.

The Highland Maya

Like the Navajo, the Highland Maya are a dramatic, troubled, and troubling people. They dwell in a dramatic setting; in Chiapas (Mexico's southernmost state) the shaggy crests that dominate the horizon and the eye give the sense of a natural order charged with power and authority, discontinuous from human purposes and projects. The native feels personally related to this land or to parts of it, but the foreign observer feels small and finds little that is reciprocal, that extends him, in these somber hills.

Like the Navajo and unlike the Lowland Maya, the Highland Maya have known rebellion followed by bloody defeat. At about the time when the Navajo were being "pacified" by the United States cavalry, the Highland Maya rose up under messianic leaders in rebellion against the Ladino, Spanish-descended landholders of their region. Like the Navajo, the Maya were crushed, and their defeat left them open to heavy oppression, including retributive massacres, by the Ladinos. Like the Navajo, the Maya learned that warlike action against the superior power brought down disaster upon their own heads and left the enemy relatively unscathed. Like the Navajo, they turned their warrior's rage inward, where it took the form of civil strife: witchcraft accusations, ambushes by feuding families, and domestic violence—all sponsored and rationalized by heavy drinking. And, while the Highland Maya are now self-victimized, their defeats have also left them politically and economically disadvantaged. Their rugged lands barely support them, and they are treated like serfs in the sugar *fincas*, the plantations, where they seek wage work. In their own region, the Highland Maya are generally scorned by the Ladino merchants and administrators, who use them as servants and live off their labor. In effect, Chiapas is the Mississippi of modern Mexico, and it is only recently that the Mexican Revolution of 1910–16 has manifested itself among them, in the form of schools, clinics, and other social services.

As a consequence of their exploitation, the Highlanders, particularly in Ladino-dominated towns, display a mixture of fright and uneasy arrogance. The men most often appear defiant, and the women tend to be in a state of near-panic, but both men and women move through the streets of San Cristóbal de las Casas (the major commercial town) like reconnoitering patrols behind enemy lines. They hug the walls, they stoop over (as if to make small targets), and they move with a quick, nervous, shuffling gait. Like infantrymen moving from cover to cover under sniper fire, they cross streets doubled over and at a half run.

In the Yucatan, among the Lowland Maya, one quickly senses, despite the alienness of native dress, custom, and language, a comprehensible "Western" ego, whose keynotes are delay of impulse in thought or action, and rationality. But Chiapas offers an "Asiatic" experience; Western life and training is no preparation for the textures and nuances of the Highland Maya character. One senses a "medieval" quality in the life of Chiapas: a heavy, fearful, and hungry note, reminiscent of the heavy-jawed peasants of a Brueghel painting. Thus, while the Highland Maya are distinct from the Navajo in terms of the norms that govern public, day-to-day intercourse, they share with them a primitive emotional makeup that relies on projection. As with the Navajo, the unseen world beyond the immediate social horizon is charged with menace, with the destructive power of witches and *naguales*, the animal familiars that do their bidding.

But if men are not to despair completely, malign power must be balanced by good power. The Navajo rely on their medicine men to bear the brunt of bad power and to oppose it with the good power inherent in their chants. The Highland Maya rely on a folk hierarchy of authority, made up of judges and attendant policemen who maintain civil order, and on the *cargo* (literally, "charge" or "responsibility") holders, who govern the church and organize its festivals, to maintain the necessary good relations with God. The most powerful judges, the *autoridades*, and the *cargo* holders tend to be old, and all elders, whatever their degree of prestige, are automatically shown elaborate deference by younger men. Approaching such a one, the younger man removes his hat and bows his head, waiting for the elder's touch of blessing to release him from his immobilized, abased condition.

This is more than token respect; the Highland Maya actively seek out, "create," and supplicate powerful authorities—whether these be Ladino employers (*padrones*), gringo anthropologists, or the town judges. The last are usually older villagers of some affluence and standing in the town who have been named to their positions by the *principales*, the established leaders of the town. The judges seem to accept their rank in a matter-of-fact way; yet their self-perception does not always match the view held by their fellow townsmen. When a citizen is in trouble or has a case to plead, he approaches the judges— even when the latter are drunk—with supplicatory, pleading, and humble gestures. He seems to see in them the omnipotent captains of his fate. In the Navajo case, the old men cast a shield of good power to blunt the effects of bad power in malign hands, but the Highland Maya seem to fear themselves and their own capacity for violence; they look to the older judges as censoring, punitive figures who either chastise them for their violence or set limits on it.

But like most holders of power, the older men can become vessels of the very evil that they punish in others. Navajo medicine men are often suspected of practicing the same witchcraft that they fight (it is said of them that they know too much about the "bad" side), and the Highland Maya believe that the *naguales* of the old men forgather at night on the mountaintops, to judge the souls of the villagers. The coven is an eerie extension of the old man's court and old men's justice. Thus, in times of famine or epidemic the elders are often held responsible and may be at risk. The motto of the Highland Maya elder is: "Quedo callado" ("I stay quiet"; in effect, "I am careful").

Besides meeting my research criteria for gerontocracy, the Highland Maya also meet the criteria for generational continuity. Marginal, preliterate agriculturalists, the younger men of this generation live much as their fathers did—cultivating the *milpa* (cornfield) and doing occasional wage work in the sugar plantations of the "hot country" (the tropical lowlands along the Pacific coast). Because Federal schools have recently been introduced in the Indian villages, the current generation will grow up differently from their parents, at least in terms of literacy. But for the generations represented in our sample, essential continuity of experience exists in all major spheres, from fathers to sons.

The Lowland Maya

Although they share a common ethnic and linguistic heritage, the Lowland and Highland Maya are vastly different in most important respects, to the extent that the Highland Maya resemble the Navajo more than they do their Lowland cousins. The differences between the Lowland and Highland countrysides reflect a broad range of differences in other domains. Chiapas is built on verticals, but in the Yucatan the eye is trained to flatness and lack of variation, constantly pulled horizontally along the scraggly but generally unbroken line of the bush, over the regiments of henequen plants, and along the straight extension of the road ahead. There is little opposition of line, color, or form, of verticals to oppose the horizontals, of clearings in the dusty uniformity of the bush, of stream or lake as contrast to the chalky texture of limestone and soil.

In the village, one finds a world whose major events take place at or below eye level. Except for the church, no structure asserts itself with much distinctness, individuality, or angularity against the sky

31

and the land. Most buildings are alike, from the ground plans within, to the conical straw roofs, to the whitewashed stone walls. Within the town, the roads meander to avoid an outcropping of limestone here, a tree there. They do not cut through the landscape but rather accent it; at their margins footpaths merge, through gradations of underbrush, into the surrounding bush. The domestic habitat is similarly mingled with the terrain; neither the exterior nor the interior of the typical house is sharply distinguished from the natural surroundings. The two realms, the natural and the domestic, are interpenetrated: The house floor is of packed dirt; the roof is of thatch; piles of corn, fodder, and wood at hearthside represent the world of bush and *milpa*. Chickens, dogs, and even pigs move freely from outdoors to indoors and out again, in their search for scraps.

There is a lack of distinction between nature and house, between house and house, and between person and person. Compared to the Navajo or the Highland Maya, dress is not used by the Yucatecan to portray individual themes or differences. The men's costume, simple and practical, varies relatively little from *campesino* (peasant) to *campesino*; the women's *huipils* (shifts) are clean, simple (save for touches of embroidery), and similarly invariant. The Lowlanders are definitely not expressive through their dress, in the manner of Indians from other groups and from other regions. Both the natural and the man-created world suggest uniformity, homogeneity, and the blending of agents that, in other settings, might manifest themselves in contrast and opposition.

But the Lowland Maya are not passive; they are modest and controlled rather than featureless. They control themselves, but they also exercise significant control in their own land, though without cutting themselves off from the larger national life. The Lowland Maya participated in the Mexican Revolution, and they are accorded some respect by the Ladinos of their region. A non-Indian cannot expect to be elected governor of Yucatan unless he speaks fluent Maya.

In some measure, Lowland Mayan autonomy depends on Yucatan geography. While it is part of the Mexican land mass, the Yucatan is de facto an island, connected to the mainland by only a few highways, otherwise cut off by swamps and mountain ranges. Within their enclave, the Lowland Maya, a generally stubborn people, have managed to imprint their lifeways on the land and to maintain them in the face of temptations toward acculturation and modernization.

The Yucatan is split between two zones: the more acculturated sisal or hemp-growing zone, where livelihood is based on wage work in the henequen plantations, and the maize-growing zone, which relies on subsistence agriculture and is more traditional in its customs. It is particularly in this latter zone that we see the continuity of tradition from generation to generation of *milperos* (corn-growers). The *campesino* fathers do with their sons as was done by their fathers: When boys are about 10 years of age, the fathers begin taking them to the *milpa*. The sons become *campesinos* long before they can even begin to consider a different life work; they are *milperos*, corn-growers, in the same sense that they are their fathers' sons, and the generational continuity that the Maya value is assured.

Unlike the North American Navajo or Highland Maya societies, the Lowland Maya do not maintain a gerontocracy: Although Mayan men can live to a ripe and respected old age, they do not routinely hold positions of social control. The authority ladder of Mayan villages is occupied mainly by younger men, and these for the most part busy themselves with practical affairs: the electrification of the village, the financing of such projects, and the relations between the village and the non-Mayan world. The younger men hold these positions because of their literacy and their knowledge of modern Mexico, but they are not treated with any particular deference, and they are not called upon to decide the fate of individual villagers. Likewise, the older men do not command any great mystique, but they are for the most part cherished by their families. Their sense of security does not seem to rest on their ability to control respect, fear, or ritual power, but on an alternate base—the sense of love and obligation that they receive from their younger kinsmen. Typically, when asked about their particular sources of well-being and security, old men respond, "I am content because I have good sons who maintain me well in my old age." And on their side, the younger men say, "My father gave me life and taught me to work; and so it is my obligation to care for him when he can no longer work."

Despite the lack of formal gerontocracy, the Mayan elders are in a better psychological position than are the elders of Kansas City. The vulnerability of elders in the United States is increased by family dispersion and by their own inner rejection of dependency. But the old Mayan thrives on freely offered filial support. He does not prevail because of his power to control other men, nature, or magic; he prevails because his security rests on good human relations. The grown children see in the aged parent, despite his wrinkles, the person who raised them when they were children and who now merits and needs their care. And in the Yucatan case, since it is prompted by love rather

33

than by fear, that care is freely given. The Yucatan elders are the beneficiaries of the internal controls that they installed, early on, in their now reliable children. But whether he is a ruler or a beneficiary, the Lowland Maya elder has a position of special advantage.

The Druze of Galilee and the Golan Heights

The Druze of Galilean Israel and the Golan Heights were the group I studied most extensively; I believe that I am the first psychologist to have been given their trust. The Druze are perhaps the best group in which to test the hypothesis that the later-life move toward a milder orientation is an internally rather than an externally stimulated event, based on genetic factors rather than on social deprivation. For while they meet the criteria for social uniqueness, for traditionalism, and for generational continuity, the Druze are also a paradigmatic gerontocracy; the elders are the favored keepers of the secret religious doctrine that defines, individually and collectively, the identity of the Druze.

In modern times, the Druze are concentrated mainly in Syria, Lebanon, and Israel. Sited for defense, their villages top the hills on both sides of the Jordan rift, including, within Israeli holdings, the Galilean ridges, as well as the volcanic plateaus of the Golan Heights. I worked in a variety of Druze communities that defined a folk-urban continuum, from the village of Daliat-Il-Karmil (exurban to the Jewish port of Haifa), to the most traditional enclaves on the slopes of Mount Meron (Jebel Jermak) in Israel, and snow-capped Hermon on the Golan Heights. In Israeli Galilee, the Druze villages sit on rocky, pine-forested hills threaded by winding roads, overlooking vineyards and olive groves. On the eastern side of the Jordan rift, the Golan hills rise to broad steppes that furnish pasturage for goats and rich volcanic soil for the Druze apple orchards and wheat fields. This eastern range of the Druze is a land of high plains, lifting toward distant ridges. Asia—the Asia of horsemen and warrior herdsmen—begins here.

The reasons for the unique power of the Druze elders are to be found in the equally unique circumstances of the group's history. The Druze are not distinguished by their ethnicity (which is Arab) or by their language (which is Arabic), but by their special, heretical variant of Islamic religion. The Druze creed was first promulgated by a Persian known as Darzi (The Tailor) and is thought to be compounded of Islamic, Hellenic, and Zoroastrian elements. A messianic religion, it initially attracted many adherents, known as the Darazeen, across the Middle East, including Egypt. The Druze were condemned as heretics, and the militant resistance to their surgent faith turned them into a persecuted minority, a condition that has persisted in varying degrees and with varying consequences, for some 800 years. They have survived in the Levant by virtue of a potent mix of skills and virtues. As the Lebanese Christian militias and the United States Marines learned in the 1984 fighting around Beirut, the Druze are and have been, throughout their history, very formidable fighters; they have often contributed, as foot troops and as military commanders, to the martial power of the host countries. In effect, the Druze offer their young men to be soldiers for the majority power, in order to guarantee the social security of their community. (This is the deal that they have worked out with modern Israel, where the Druze are the only significant non-Jewish minority whose young men are drafted into the Israeli defense forces.)

Coupled with their martial prowess, the Druze have a well-deserved reputation for political acuity—vital to any embattled minority—and for going with the winners. They have a trained capacity to operate in the jungle politics of the Middle East and to sense who is going to come out on top in regional struggles. Thus, in 1948, when most observers thought that the infant Israeli state would be snuffed out, the Palestinian Druze recognized Israel's viability and threw in their lot with the Jews, emerging from the war as a privileged minority within the Hebrew state. This quality of political shrewdness is summed up in a Druze joke: "The Druze flag has two ends. Why? One end is to wave when the conqueror arrives; the other end is to jab in his ass as he leaves."

This is not to say that the Druze are essentially a mercenary or opportunistic people; they bend and accommodate not out of fear or cupidity but in the service of a higher goal—the preservation of their religion and traditional lifeways. They will give Caesar what he requires, so long as he does not violate the basic tenets of Druze honor and religion. For the Druze have another saying: "The Druze are like a brass plate—strike one corner and the whole will resound." If accommodation does not succeed in its purpose of protecting the Druze priesthood or womanhood, then the Druze will, as one man, go to war.

35

But the Druze have also survived through disguise, through dis-sembling their problematic religion. Much like the Marrano Jews of Spain, who hid their Hebraic faith from their Catholic neighbors dur-ing the Inquisition, the Druze often had to keep their religion a secret from vigilant Muslim neighbors. Among other strategies, they main-tained security by not telling their children that they were Druze until they were old enough to be trusted with that dangerous secret. Eventually, this cautious practice acquired a momentum of its own, and it persists even now that the oppression has lessened. Druze in-dividuals are entrusted with the secrets of the religion after they have lived an exemplary life. Accordingly, entry into the inner ranks of Druze religion is offered mainly to older individuals, those who have successfully raised their families and made valued contributions to their communities. The young men are *jahil* (unknowing); for the most part, it is postparental older men who enter the privileged ranks of those who know, the *aqil* (pronounced ah-kell), and are given special access to the secrets of the Druze religion, and their own special, scribe-written copy of the arcane text. Accepting the invitation to become *aqil*, the Druze man adopts special garb (including the *lhafi*, a white hat topped with red cloth), shaves his head, grows a beard, and renounces tobacco, spirits, and loose behavior. Twice a day he goes to the *hilweh*, the prayer house, with other *aqil*, there to beseech the grace of Allah for his community, his family, and himself.

Inner psychological changes accompany the outer, behavioral changes. Typically, the older *aqil* seems to feel himself reborn as a child or even as a special extension of Allah's will and grace in the world. This identity provides powerful personal meanings that make possible important renunciations; besides giving up the usual sensual indulgences, the *aqil* accepts an even more stringent control over his inner life. The elder *aqil* is expected to devote himself to good and pious thoughts, to avoid lustful reflections, and to forget the errors and stupidities of his life before he was introduced to true knowledge.[2]

Because of his sacrifices in its service, the *aqil* exemplifies the religion that defines and identifies the Druze. The *aqil* experiences himself and his people as identical, and he sees himself as refashioned to the requirements and measure of Allah. In a quite direct and literal way, the *aqil* becomes the joint that binds and interfuses the two great entities—the Druze people and their god—making them real to each other.

But the *aqil* do not devote themselves exclusively to mediating the relationship between the community and Allah; they are also a powerful force in community affairs and are consulted at important political junctures. But whether they are directly involved in particular decisions, the heavy, sometimes reproving, and always dignified and portentous presence of the *aqil* pervades all aspects of Druze life. These elders provide the moral weather for the community, the context for all decisions, whether these bear on sacred or secular affairs. Of the four cultures that I have studied, the Druze are the quintessential gerontocracy.

Summarizing the various roles and powers of the aged within several peoples, we see both differences and similarities. The Druze *aqil* provide a living link to Allah, and the older Navajo medicine men serve similar functions, calling up through their rituals the people's mythic beginnings and the good power that moved the mythic events. But the Druze *aqil* provide a constant, daily reminder of Allah's stern presence, while the Navajo call up their gods only sporadically and mainly at times of trouble. By contrast to them both, the Highland Maya judges and old *autoridades* do not directly represent the gods but are imposing, even frightening, in themselves. It is only the Lowland Maya elders who maintain their position through love, rather than fear or awe. Nevertheless, whether they represent a channel to *tabu* forces (as in the Navajo and Druze cases) or are themselves the wellsprings of *tabu* power (as in the Highland Maya case), in three out of the four study sites the old men control supernatural, even awesome powers, and represent, in external form, the inhibitory, controlling functions that underwrite individual propriety and a reasonable degree of social order.

Let us turn now to a brief discussion of the field methods and instruments that were applied at the sites, as well as the kinds of data that they generated.

Naturalistic Methods in Cross-Cultural Field Research

In our society, the process of aging and the aged who represent that process are both stigmatized. It is the fate of stigmatized objects to spread their infection to everything that bears on or derives from them. Gerontologists are attracted to the study of aging, but they are also repelled by the stigma, by the fear that it will attach to them, in the same way that the aura of death attaches to undertakers: "What kind of creep studies aging?"

Accordingly, in order to fend off the stigma, gerontologists have concentrated on the "clean," rigorous, ultimately "masculine" aspects of aging phenomena: numbers, methods, the cosmetics of "hard" science. The result has been a premature rush to quantification, a focus on method and rigor that serves, incidentally, to distract us from the dangerous excitements of our field. As a result, we have bypassed the naturalistic, data-generating phase that is vital to the development of a true social science. Instead, we have elegant cross-sequential grids that wait for data but that have also in large part predetermined the nature of the data they will accommodate.

We now recognize that the gerontologist's search for scientific purity has led to a dubious trade-off: security instead of surprise. We go to the field with blinders on, looking for the kinds of precoded, quantifiable data that will be acceptable to a canned computer program.

But the house of science has many rooms. It does not only call for disciplined testing of a priori hypotheses; it also has an exploratory aspect, one that leads to the generation of new data. We test established hypotheses by the rigorous management of precoded "data" whose meaning is already fixed; but new data are generated only via spontaneous, naturalistic approaches. Reversing the usual research canons, these call for free-floating rather than focused attention and an openness to unpredicted regularities.

We begin naturalistic investigations by accepting the blunt fact of nature, that it has its own intrinsic order, independent of our theories about it, and that, if we stand ready to abandon our preconceptions, this order may declare itself to us, in its own terms. We should go to the field with initial predictions and with the questions that devolve from them, but we should always be aware that these are kamikaze ideas: They serve us best by dying, but only after they have revealed their inadequacy, after they have highlighted the unforeseen events and regularities for which they cannot account. As our investigation develops, the key, recurrent events that are the signature of these regularities, even though they have previously gone unnoticed, become our data (and may even become data for those who would use them to refute our ideas).[3]

Discovering the Naturalistic Interview

I started my field work in a conventional manner, employing a standard interview administered in a standard fashion. I used my subjects as informants on their society, rather than on themselves. I asked them about the customs and attitudes that governed the relations between young and old in Highland Maya society. I soon discovered that scientific respectability was bought at the price of boredom and small results. After the fourth interview, I could pretty well predict the rest. In response to my interview questions concerning the fate of the aged, Highland Maya men insisted that sons respected fathers and that fathers instructed their sons most scrupulously in all things. As a clinical psychologist, I knew that this rosy picture represented a gross simplification of complex matters but could think of no method that would deliver up data from the informant's proactive and surgent levels. Besides being stymied by conventional methods, I also had logistic problems. In the usual manner of the Highland Maya, my interpreter became soddenly drunk. In addition, my interpreter's buddies, most of whom had already been interviewed, often interrupted the work in their search for drinking money. On one occasion, as I was trying to counsel my interpreter into sobriety, one of his friends lurched in to demand ten pesos. To get rid of him, I complied with his request, but he reappeared shortly, with two quarts of *trago* (hard liquor) and the rest of the drinking cooperative. Understandably grateful, they called me *papacito* (little father), explaining that the man who bought them *trago* was their father and they were my sons. Entering into the drama, I replied that I was an unfortunate old man to have a bunch of drunken bums for sons, and who would maintain me in my old age? Delighted, they explained that the father should beat his son with his belt to curb the drinking: "When the father beats the sons they fear the belt more than they want the *trago*, so they do not seek for *trago*." When I did not play the required heavy father role, my unbeaten sons swung from deference to demand, insisting that I find them women: "Your sons are grown up now and they have no wives; you must find women for us!" Presenting his bowed head, one said, "You grew my head, now you must find me a woman." Adding fuel to the fire, I replied that I was obliged to give them only a mother, not wives. "No, no," they said, "the mother belongs to you; now you must get us women of our own!" Another man muttered

39

that fathers who didn't get wives for their sons got killed.

In one sense, Freud's thesis in *Totem and Taboo* (1913) was being reenacted: I was the father who possessed all the women, and my former informants were the band of rebellious brothers, demanding their sexual rights. The next step could be to kill me, eat me, and deify me. I was spared this lurid fate because my "sons" actually had a different agenda: My research subjects were dramatizing the real state of affairs between aging fathers and adult sons in their village of Aguacatenango. The idealized version, reported in response to survey interviews, proposed that aging fathers and sons were bonded together by mutual respect and nurturance, but the psychodrama staged by my informants showed otherwise. Respect for fathers was founded on *force majeure*, and paternal instruction was delivered to sons by means of a belt buckle. My informants could not put this truth into words for me, but they could, in more concrete fashion, act it out.

The structured interview had tapped only the sculpted norms for aging fathers and adult sons. By switching media my informants showed me the species context of this relationship: the universal, surgent drama of the Oedipus complex.

The Naturalistic Interview: Building Rapport

Via Mayan media, I finally got the message. Incidents like this taught me that standard interview procedures and protocols are based on an egocentric illusion, namely, that the investigator's fixed procedures will "standardize" the entire situation, including the subject's understanding of the interview itself. But as I soon discovered, even within high-consensus village societies, individual respondents would have grossly different understandings of my purposes and of the meaning of the interview, ranging from the suspicion that I was a government agent to the delusion that I had been sent by God to chronicle the history of the village before the imminent destruction of the world. Evidently then, the "apperceptive mass," the preformed system of related ideas and sentiments that respondents brought to the interview was different in each case and could grossly influence, in untraceable ways, the degree of rapport and the quality of the data.

Clearly, the standard procedures would neither reveal the subject's apperceptive mass nor reduce its effects; it would only leave these biasing distortions forever planted in the data. The subjects' preoccupations can be addressed only in their own terms and in ways tailored to the subjects' actual concerns. The standard interview condition is not characterized by concretely similar procedures from case to case; rather, the standard condition is one of rapport—the subject wants to talk to you in depth about matters that are important to him as a person and to you as a scientist. But before such sympathy can be achieved, the subject's distorting apperceptions must be revealed, and he must be given a reasonable and reassuring experience in terms of them. Thus the standard situation of rapport is in each case reached via different and nonstandard approaches.

The search for good rapport started before the interview proper, with the entry into the community itself and the selection of co-workers there. At all sites, I sought out and explained my purposes to community leaders who could vouch for me and legitimize me to their people. For the most part, these community leaders either worked with me as interpreters or selected interpreters from among their own protégés. Consequently, I was legitimized either by some authority or by those who inherited some of the leader's influence. We did not try to select a statistically orthodox random sample, as the informants selected by such pure but unrealistic designs usually proved to be sick, away from their village on wage work, off at some distant tillage, or in jail. Instead, in each community we tried to interview all possible subjects within the approximate age range 35–80. Because we tried to be respectful of the individual subject, our legitimacy usually grew, to the extent that potential subjects sometimes asked to be interviewed: "Why do you avoid my house?"

The Invited Interview

Continuing the quest for the kind of rapport that would illuminate all the Linton levels, I routinely opened the interview by inviting the respondent to question me, on the grounds that one does not speak intimately with a stranger until one knows something about him. Since the respondent's questions are an index to his preoccupations, about the interview and other matters, this invited interview often provided my best data, as well as my surest access to the respondent's inner world. Most important, the invited interview turned out to be an effective means for bringing the apperceptive mass into the open and for neutralizing it. Thus the respondent's questions to me often provided valuable clues about his covert concerns and suspicions, and gave me an opportunity to counter them: "You suspect that I am a government agent and that I have come to find out if your taxes should be raised. That is a reasonable suspicion, and I am not going to argue with you about it. All I ask is that you go through the interview with me and decide for yourself if I am asking the kinds of questions that a government agent would ask." Reassurance was usually given by openly taking up the subject's suspicions, instead of arguing against them.

At the outset, it was also important to demonstrate, sometimes in unconventional ways, that I was genuinely interested in the person and not in the public relations that he prepared for foreign consumption. For example, Druze *aqil* typically welcomed me into their homes with elaborate protestations of cooperation but then spent the opening moments recounting the history of the Druze rather than their own personal story. In part, this confounding of person with people reflected the identification of the older Druze with his community; as he saw it, personal history and collective history were conjoint events. But it also reflected a defense against entering—and revealing—the private world of surgent emotions and associated vulnerabilities. After I objected, saying, "I can read about the history of the Druze in books, but I can learn your history only from you," the older Druze subjects would usually be more forthcoming. Now convinced of my interest, they might then reveal their private relation to their own history, to their people, and to Allah.

The Language of the Stranger

In all cases, the contact language was other than the local tongue: My American Indian guides interpreted from Navajo to English; the Mayan contact language was Spanish; and among the Druze it was again English. Purists will argue that ethnographic interviews should be conducted in the native dialect, but I hold that the goals of naturalistic interviewing are best served by a stranger who speaks in a foreign tongue.

The goal of the naturalistic interview is not sentimental; it is not to become buddies with the subject but to elicit data that are foreign to both the investigator and the subject himself—the data of private wishes, fantasies, and constructions, the language of the preconscious mind, even the surgent unconscious. The alien, the stranger who does not speak the local language, is a fitting metaphor and reciprocal of that which is alien within ourselves. The stranger is so defined by the fact that he is not part of our consequential social system; confiding in him, we run no risk of rejection. The introduction of a foreign language and a foreign interviewer gets the intended idea across: that the interview is taking place outside of the usual social space and that unconventional matters, foreign to the usual social discourse, can therefore be explored. Indeed, the whole thrust of the interview methods was meant to confirm the subject's experience of me as an interested stranger, a naive and even stupid man, but one who, out of genuine interest, insisted on being instructed: "You say that your life has been corn and beans. Another Maya would understand what you meant. But I am not a Maya, I am a gringo, and so you must explain to me about these things. Now: What about corn, and what about beans?"

Old men are often garrulous. They love to talk, and a host of studies on the uses of reminiscence suggest that it is generally good for old people to recount their personal histories to an interested listener. Through reminiscence, older persons may creatively revise their past to conform to some sustaining personal myth; by finding or constructing the threads that connect the current self with some past, idealized self, they can overcome the devastating experience of discontinuity and self-alienation. Accordingly, once it had been demonstrated that I was truly an interested stranger, informants often used our interview as an occasion—in many cases, for the first time—to create, to construct, their personal history. Although my approach did not always ignite a significant exchange, many subjects, across study sites, testified that the interview had been a unique and pleasant experience: "Nobody ever asked me these questions before."

In practice, I shaped the interview to serve two major priorities: The interview should elicit the personal construction of individual lives, and the interview should elicit data that would contribute to the testing of hypotheses and to the development of a comparative theory of aging. In keeping with these distinct goals, the interviews on the one hand covered a standard agenda of issues: early memories, including recollections of parents and siblings; major events in the course of the adult life span; medical status; current life situation, including sources of gratification, sources of displeasure, and remedies against displeasure; dreams and reveries. On the other hand, in order to serve our exploratory goals, these questions were not covered in any fixed order, but only as they matched the priorities of the interviewee; his subjective standards were allowed to direct the course of the investigation. In the course of the interview, the informant would presumably teach me how to question him; and I would teach him to be an informant on his life—his own father, his own son—rather than the general norms of his society. As we explored the informant's private, subjective relationship to the norms, we opened up, for exploration, the private stances, the occulted motives, that distinguish subjects from each other, even within high-consensus traditional societies. Paradoxically, by concentrating on Linton's third level, the unique man, we may invite the disclosure of Linton's first level, the voice of the surgent unconscious, the voice of universal man, the voice of developmental process.

The Projective Tests

The naturalistic interview is one route into the psychic underworld; the analysis of private fantasy and imagery is another. The first, surgent tremors of a developmental sequence will announce themselves in dreams and reveries long before they filter into conscious thought, discourse, and behavior. Therefore, in order to further test and explore the developmental propositions, we topped off the interview with the administration of projective tests.

Indeed, besides generating data in its own right, the interview also serves another major task: preparing for the administration of the unfamiliar, sometimes threatening, Thematic Apperception Test cards. As the subject interprets the ambiguous TAT scenes, he not only describes the cards in their objective sense but also reveals his subjective, often ambivalent relationship to the essential affective issue or conflict represented in the stimulus card. Subjects usually recognize, if preconsciously, that they are not merely describing a fixed stimulus but that in so doing they are also reporting on some aspect of their inner life. Thus the projective tests pose a specific threat: The subject is asked to surrender control of his public relations, without knowing what the examiner is making of his self-disclosures. The successful interview serves to generate a special degree of rapport, the precondition for successful projective testing. That is, in the course of a depth interview, in which private concerns are aired, the subject can learn that the interviewer will not be put off by the informant's confidences, no matter how personally troubling these may be. Presumably, the resulting sense of assurance will render the projective testing less stressful and lead to more open, unguarded responses. Under these conditions, the TAT can become a continuation of the interview by other means, and the royal road to the surgent unconscious life of the elder. The TAT was administered to all sighted respondents, following the completion of the interview. The battery consisted of certain cards from the standard Murray TAT set, those which—like the Rope Climber (discussed in the next chapter)—depicted no particular culture or type of person, as well as some cards that were either drawn up or altered to fit the local norms of particular societies, while conveying a standard meaning across study sites.

Thus, in its Indian version, the card of a boy standing on a desert mesa shows him wearing a loincloth and moccasins, but in the Druze version, he wears long pants and is barefoot. Through these alternations we preserve the card's stimulus meaning—"a watchful boy on a high place"—without introducing, for a particular culture, the surplus and distracting idea: "This is someone strange to us."

The TAT administration was prefaced by the instructions: "Tell me a story about what you see here, with a beginning, middle, and end." However, many literal-minded peasants did not grasp the idea of the card as providing the occasion for an exercise of imagination; in their mind, the stimulus and the thought that it excited were one and the same. A particular card could contain only one fixed story, and it was their job to discover that singular, "correct" meaning. Frustrated in this quest, they might say: "Señor, it is not written here what the story could be"; or "I am a poor man; I have no education." In such cases I amended the original instructions to: "Tell me what you see here," and then, starting from the subject's initial interpretation, I would use nondirective probes to draw out a full story: "You say that these two men appear angry—what could they be angry about?"

I describe the analytic schema for the TAT and interview data in succeeding chapters, at those points in the text where these data are reviewed in detail. However, as with the data collection, the data analysis also followed from naturalistic principles. Thus the working assumption guiding the TAT analysis was that each response represented a motivated communication, an answer that summed up the respondent's subjective stance toward the basic conflict or emotional issue portrayed, however latently, in the card. And as the motivated aspect of the subject's communication is clarified, the voice of the surgent unconscious is amplified, as it stirs toward new growth and development.

The theoretical approach of our study will be clarified as I present, in the next chapters, the analysis of the data generated by the interview and by particular TAT cards across the range of research communities.

3

THE LIFE COURSE
OF MALE AGGRESSION:
FANTASY EXPRESSIONS

Young Men, in the Conduct, and Manage of Actions,
Embrace more than they can Hold, Stir more than
they can Quiet; Fly to the End without Consideration
of the Means and Degrees; Pursue some few
Principles, which they have chanced upon absurdly;
Care not to Innovate, which draws unknown
Inconveniences: Use extreme Remedies at first; And,
that which doubleth all Errors, will not acknowledge
or retract them; Like an unready Horse, that will
neither Stop, nor Turn. *Men of Age*, Object too
much, Consult too long, Adventure too little, Repent
too soon, and seldom drive Business home to the full
Period; But content themselves with a Mediocrity of
Success.

—SIR FRANCIS BACON, 1561–1626

Psychological changes, potentially developmental in nature, come
about in later life, and these changes affect men and women differently
at all levels of personality organization: species, psychosocial, and
idiosyncratic. Since the most profound and universal redeployments
presumably take place at the genetic, species levels, we will first consider the deep emotional, or instinctual, life. (In this and the two
following chapters, the focus is on men. Women, for whom we have
less primary data, will be considered at length, in their own right,
beginning with chapter 6.)

As indicated in the last chapter, the best methods available to us
for the study of depth psychology are the projective techniques, the
"strange," unfamiliar stimuli that invite respondents to externalize

and explore the estranged, unconscious parts of themselves. I will use the data developed by these instruments to begin the exploration of deep human motives, of love and hate—of eros, the binding force, and of thanatos, the disuniting, destructive force in human affairs.

The Life Course of Male Aggression

The bloody history of humankind testifies to the singular and enduring power of aggression as a major theme in individual and collective existence, particularly in the life of males. From the species or procreative perspective, men are more expendable than women: One man can inseminate many women, but women bear their children one at a time. There is, in effect, a surplus of men, who are accordingly told off for the necessary high-risk, high-casualty tasks on the borders of the community. "When it comes to slaughter, you do not send your daughter" is one of the most common rules of human society. Because of its central place in men's psychic and social life, let us start by investigating the life course of male aggression.

Like love, its counterpart and adversary, male aggression serves individual and group survival. It has its ultimate basis in the hormones, in testosterone, and in the special sex-linked neural structures that are selectively responsive to neurohormonal stimulation (for example, by testosterone). In its most primitive and direct expression, aggression underwrites individual survival under harsh and primitive conditions. In its basic form, aggression is expressed as the readiness for fight or flight: the readiness to eliminate obstacles to survival or pleasure, and the readiness to flee when the battle is lost. But in the course of individual development, aggression is tamed, socialized, put at the service of collective survival—of the family, of the peer group, of the nation. As maturation proceeds, individual aggression is split off from the service of individual needs, individual fears, and individual greed, and put to the service of "the other," the collective entities whose survival is given priority over the survival of the individual. In its "sculpted" manifestations, aggression is split from its original goals, and instead emerges in forms of behavior that promote social survival, as well as individual security and individual advancement. Thus athletic competition, political leadership, bargaining, sharp analytic thinking, and jurisprudence give rein to individual aggression, while simultaneously serving the needs of the social order. Although aggression originates as an antisocial impulse, it normally becomes prosocial, pervading and even underwriting most aspects of orderly social life, particularly those dominated by men.

Despite its central place in men's psychic and social life, there has been no investigation into the life career of aggression. The young man's deployment of aggressive powers into ambition, into self-defense, and into the defense of family and nation is taken as the model for the entire male adult life cycle, even in the later years. However, my earliest investigations with data from the United States cast doubt on this untested assumption, suggesting instead that the male endowment of aggression is not fixed over the entire life span. Instead, in those samples, it declined with age, sinking below the mean level of female aggression. On the assumption that the American case was not unique and that these findings pointed to a species rather than a cohort, or age groups, phenomenon, I turned to the cross-cultural laboratory to test the following hypothesis: The decline in male aggression is a universal phenomenon, not parochial to Kansas City.

Accordingly, in this chapter I will present the data generated by those TAT cards, used cross-culturally, that bear most directly on the degree, the targeting, and the management of male aggression. The data from two cards will be presented: the Rope Climber, which graphically depicts a vigorous, powerful male body, and the Heterosexual Conflict card, which depicts the expression of male aggression in a domestic, female-centered setting.

The Body's Assertions: The Rope Climber Card

This card, part of the original TAT battery, was used in its original Murray-designed form at all sites. It depicts a muscular man, possibly smiling and possibly unclothed, looking upward and holding onto a rope in midair. The card is ambiguous as to whether the climber is ascending, descending, or hanging immobile on the rope. No background is clearly indicated, although a corner, at which two walls meet behind the climber, is suggested. The climber is alone, and he may be nude; the card gives few hints of a social universe. The only artifacts that betoken a collective, social world beyond the climber are the rope and the hint of a wall. For the most part, the stimulus space is filled with the representation of raw vigor. Accordingly, the latent stimulus demand (LSD) of the card has to do with the untamed, ultimately asocial wellsprings of male aggression. In effect, the card is a representation of the aggressive side of man's id, the life of the untamed impulses. Further, in its more subjective sense, the card not only pictures these matters but asks questions about them. A continuation of the interview by other means, the card asks the respondent, "What is your conception of dangerous power, of aggression; what is your private relationship to your own assertive energy?"

This phrasing of the card's LSD is based on a transcultural consensus: Approximately two-thirds of all respondents, regardless of culture, agreed that the rope climber is pushed by his own assertive energies toward productive or destructive behaviors. The Highland Maya are the only group in which less than half of the respondents see the rope climber as a self-motivated, energetic figure. Normally then, the rope climber is seen to be striving energetically toward some productive goal: He is a competitor in a sporting event, he is an athlete training for some future trial, he is searching for wealth, he is harvesting a crop, he is seeking a shortcut through rough country.

This intercultural consensus concerning the vigorous, assertive nature of the rope climber gives us the right to assume that most respondents, even those who do not visualize a powerful figure, are nevertheless covertly sensitive to the LSD and are coding their subjective reaction to the issue of male power into their responses (even in those stories where the rope climber is presented as powerless or as menaced by power external to him). This analytic approach allows us to view all responses as aspects of a dialectic exchange between the respondent and the card LSD. Thus the respondent who discovers a vigorous, ascendant, or triumphant rope climber is not only giving us a reasonable version of the stimulus, he is also reporting the ease that he himself feels with the organic power encoded in the card LSD. By the same token, the respondent who identifies the rope climber as tired is not only telling us that he himself feels vitiated; in the more dynamic sense, he is revealing his fear of the vital energy thrust at him by the card: He responds to the LSD of power by telling us, defensively, that he is empty of this dangerous quality.

Each phrasing of this dialectical interchange between respondent and stimulus is reflected in a particular response theme. In order to make these motifs comparable, across cards and across subjects, I viewed each theme as meeting the criteria for a particular *Mastery Orientation*: Active, Bimodal, Passive, or Magical. Each of these positions describes a basic posture toward internal and outer experience:

direct intervention (Active Mastery); alternation between active and passive stances (Bimodal Mastery); accommodation and propitiation (Passive Mastery); or revision of sense impressions (Magical Mastery).

Working from the assumption that the response to each card expressed one or another of these orientations, I translated the mastery themes into specific criteria for analyzing and locating each story on the continuum of mastery types. The standard mastery positions are listed below, and the general criteria for psychological mastery are restated to take account of the particular stories "pulled" by the Rope Climber card.

ACTIVE MASTERY. Grouped in this category are all responses in which the hero is portrayed as vigorous and successful in competition with others, as well as all those responses in which the hero is seen to strive competently toward some significant, productive goal that he has set for himself. In all cases, he is in contention: against a human opponent, against some segment of refractory nature that must be tamed or harnessed, or against the passive, fearful, or refractory parts of himself that would retreat from the struggle. In such responses the intended product of the hero's forceful action might be victory over an opponent, a geological find, or public recognition. In the cases in which the climber strives against his own inertia or laziness, the "product" is a stronger body, achieved through disciplined exercise.

Here is an example of an openly competitive theme, from a Navajo respondent: "Seems like there's something way up high, maybe a prize, and he's going after it." In the following story (from Kansas City), the hero strives actively to produce a good experience, as well as entertainment for others: "Well, I would guess that this character is an acrobat, engaged in aerial acrobatics. I'd say he's on his way up to his performance and that he is looking over the group of spectators gathered for the performance, and it looks as though he might be happy— he doesn't look worried. He looks as though he is equipped to get the job done, mentally and physically."

QUALIFIED ACTIVE MASTERY. Under this heading fall the cautionary tales: those in which aggressive, competitive behavior puts the hero at risk and even leads to his destruction (QAM₁). Also grouped here are those stories in which the hero strives with some energy but toward rather limited goals (QAM₂). Overall, in the Qualified Active Mastery stories aggressive action is either inhibited or the point is made that it should be inhibited.

In the QAM₁ cautionary tales, the climber is at risk from the same energy that is also his resource: He has created trouble for himself by virtue of his boldness and drive. The climber reaches the top of the rope ahead of his competitors, and the rope breaks, or he is a prisoner escaping from a jail, thus inviting pursuit and punishment. Here is an American example of this genre: "Possibly, someone in a physical training class in a gym or something of that kind. The picture no doubt is of a strong, muscular-type fellow, and, from his expression, he wants everybody to know it. It looks like he is pretty high up. [How does the story end?] Maybe he has too much self-assurance." The respondent first celebrates the climber's "self-assurance," but then shows his mistrust of such boldness: One can have too much self-assurance.

But despite their qualifiedly pessimistic outcomes, stories of this sort are judged to essentially reflect an Active Mastery stance (though one that could lead to or justify overt passivity) toward the card issues: The hero is ultimately the author of his fate, however bad that fate might be.

In the QAM₂ stories, the climber moves with energy, usually toward self-selected though picayune and short-range goals. In Navajo stories of this type, the climber uses the rope as a shortcut through rough country, while non-Indian North Americans often see him as a performer who efficiently fills a minor role: He is a trapeze artist routinely going through the motions of his trade in order to "make a buck." In such versions, the climber has energy, but that energy is harnessed to small purposes and bound to minor (albeit socially useful) roles.

BIMODAL MASTERY. Two sorts of stories group under this heading: those in which a passive theme intrudes into an otherwise active story (Bi-M₁) and those in which the respondent gives two discrepant interpretations of the stimulus, one active and the other passive (Bi-M₂).

Stories of the first type, Bi-M₁, include those in which some impersonal element—for example, the rope or a slippery wall—blocks the climber's active striving. Thus (Navajo respondent): "He wants to climb way high up, but his feet keep slipping on the cliff." Here the forces that would render the climber passive and ineffectual are ascribed to the environment and are thereby removed from the climber himself. In other stories of this type, the climber moves purposefully but always toward some source of immediate pleasure or security.

The wish for comfort rather than production drives such behavior: The climber brings down fruit from a tree, or he is climbing up the mesa to reach the security of his home. Alternatively, the climber is playful and relaxed, but this playfulness has a by-product—it provides the climber with a living and his audience with amusement: "He is playing—as a clown in the circus. He is amusing the others. This is how he gains his living" (Lowland Maya respondent).

In the Bi-M₂ stories, active and passive possibilities are not integrated, nor are they combined into a coherent story. Typically, the respondent gives an active interpretation that is then countermanded by a passive one, as if the subject is sensitive to both states but sees them as mutually exclusive, not capable of being blended within the same story (or, by extension, within the same self).

PASSIVE MASTERY (SENSUAL). Grouped here are those stories that convey pleasurable, dependent, and consummatory aspects of passivity. In them, the rope climber is seen to be enjoying such pleasures as are available, with minimal effort, to a man on a rope: He is resting, while looking at a beautiful view, or (Lowland Maya), "He is Tarzan, playing on the rope like a monkey," or he is swinging idly over a body of water, ready to drop and swim. Unlike the superficially similar stories grouped under Qualified Active Mastery, the hero is not seen as moving actively to acquire some portion of pleasure; rather, he is lazily enjoying his own body, his own activities, or some pleasant aspects of the outer world. Further, the climber does not actively oppose gravity so as to accomplish his sensual purpose; rather, he is poised motionless on the rope, or he swings laterally on it (thus borrowing the forces of gravity as well as using his own strength), or he falls, again pulled by gravity, into the inviting water beneath.

PASSIVE MASTERY (CONSTRICTED). These stories either have themes of constriction and inhibition, or they are products of a prior mental and emotional constriction on the part of the respondent. Either the climber himself is seen to be apathetic and immobilized, or the story itself, whatever its theme, is meager and weak in the structural sense, that is, without elaborations, connections, plot.

In even the most vigorous stories indexed under this heading, the climber is moved from without rather than from within: Essential energy does not originate within the climber but is located in some outer threat to which he responds. The climber moves (upward or downward), but his actions are not in the service of personal goals;

he is responsive to pressures and initiatives set for him by some malign external agent (PM,). The rope climber flees to escape a fire, a beast of prey, or a human enemy. (In these cases, he has not provoked the assault. Otherwise, the story would be accommodated under Qualified Active Mastery.) In response to the card's implicit question, "What is power, and what is your relationship to it?" these respondents in effect answer, "Aggressive power is an alien, suspect quality that is not part of me; it is dangerous, and my task is to keep away from it."

More extreme versions of constriction emerge in stories (PM$_2$) that depict the rope climber as unable to move, either because of external constraints ("He is stuck in a well") or for internal reasons, such as lack of personal resource ("He is tired, hanging on; slipping down the rope"). Often, because the rope climber cannot trust his own strength, he must look to human or divine aid to get him out of his fix: "He is stuck in the well; he waits for someone to throw him a rope. Perhaps only God can save him" (Lowland Maya).

Formal rather than thematic expressions of constriction emerge in those stories told by respondents who lack the energy, the motive, or both, to go much beyond clear and obvious stimulus details (PM$_3$). Major details are accurately though minimally perceived, but the relations among them are not spelled out: "He is going up on a rope; I don't know why. [What will happen?] How can I tell? It is not written here" (Druze). Or, "How can I tell? Of course, he will have to go down the rope" (Highland Maya). Caution, apathy, stubbornness (or all of the above) may prevent these respondents from enlarging on their basic perception, but in all cases, structural constriction and thematic emptiness result.

MAGICAL MASTERY. Two sorts of stories, thematically distinct but structurally alike, are cataloged here. Basically, they feature gross and wishful misperceptions of major stimulus details, so that the story is distorted away from all reasonable, conventional possibilities, and toward extravagantly destructive or extravagantly benign interpretations.

Typical of the first, tragic variety are the "paranoid" (MM,) interpretations, those in which an important aspect of the card is misperceived in order to justify an unusual depiction of the hero's activities along the lines of murder, suicide, voyeurism, or fatal illness. In many instances, the rope itself becomes the hero's enemy rather than his resource: It strangles him, it is a snake against which he struggles, or it is a spear that skewers him from mouth to anus. Presumably these respondents are committed to seeing the world as full of threat and themselves as innocent victims.

The second class of Magical Mastery responses (MM$_2$) is formally similar to but thematically the converse of the paranoid, injustice-collecting variety (MM$_1$). Again, while one or more major stimulus elements are determinedly misperceived, in MM$_2$ stories any painful, tragic, or obscene card implications are denied rather than amplified. Thus a North American respondent, when handed the card, immediately turned it on its side so that the climber appeared to be reclining rather than upright: "He's sleeping," was the response. By a simple change in card orientation, this respondent succeeded in changing the stimulus conditions on which the card LSDs of power and effort are based. This simple but arbitrary revision of the stimulus conditions permits him to deal with issues of inertia and powerlessness, rather than more troublesome concerns involved with personal power. Such respondents hold the card responsible for their own retreat from dominance.

Other respondents do not even bother to change the stimulus conditions; instead, they happily confabulate pleasant though fantastic interpretations that have no basis in the card: The rope climber is immersed in a body of water, enjoying his swim; he is an angel in Heaven, or some other divine figure. In either case, whether he is powerless or omnipotent, the problem posed by the card LSD, the problem of raw energy, has been "magically" resolved.

All available responses to the Rope Climber card were coded, blind for age, into the above categories on the basis of their manifest content. Responses from each culture were analyzed separately. The resulting distribution of responses by age and mastery style are shown in figure 3.1, indicating an association between age and mastery type, across cultures, that is significantly beyond chance level, by a statistical test of probability (the chi-square test).

Figure 3.1 was derived by computing the percentage of responses in each mastery category for two age cohorts: men aged 35 through 49, and men aged 60 and over. A comparison of the resulting profiles highlights the nature of the differences, in terms of mastery orientation, between older and younger men. These differences are in the predicted direction, in that the youngest men achieve predominance within the Active Mastery, Qualified Active Mastery, and Bimodal Mastery cells. That is, younger men specialize in those responses that portray the rope climber as an active, assertive figure, even though

FIGURE 3.1

The Rope Climber Card: Distribution of Responses by Age and
Mastery Style (Kansas City, Navajo, Druze, and Maya)

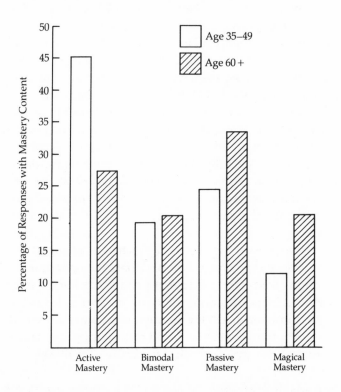

his aggression may be admixed with or countered by passive leanings (as in Bimodal Mastery). And even in their passive responses, the younger men mainly adopt the mode (PM₁) that involves the deployment of external aggression (originating in the outer world, not emanating from him) against the victimized rope climber. Thus younger respondents eschew the most dispirited versions of Passive Mastery, those in which both the climber and his environment lack power. Even for passive younger men and even though it may be outside of them, power is still an important concern.

In short, whether they use it, abuse it, or flee from it, power is the younger men's variable, as revealed by the Rope Climber card. By contrast, the older men gave few responses that could be cataloged as active mastery, suggesting the personal display of power. They are also lower than younger subjects on PM₁, the phrasing of Passive Mastery that registers the targeting of aggressive power against the self.

In addition, the old men score high on Bimodal Mastery, which captures an ambivalent, dubious stance toward personal power, and they also peak on paranoid Magical Mastery, a position that stands for alienated aggression, for power that is out of control in the world, in the self, or both. Passive Mastery (Constricted type) may be an index of anxiety, and the older men's lead over the younger men on this barometer suggests that they may become more fearful as they become more alienated from their own aggression, experiencing it as an external threat rather than an internal power resource.

As estimated by the Rope Climber card, in later life the aggressive currents leach out of those mastery positions in which power wears a human, or "natural," face. Having ebbed away from these stations, it reappears, haloed with anxiety, in paranoid Magical Mastery. In later life, male aggression takes on an inhuman and frightening form, to become a property of all that is strange and unnatural, within the world and in the self.

Despite much transcultural unanimity in the age grading of Rope Climber card themes, our inferences concerning age change are shaky, in that they are based on cross-sectional data. Thus, while it is unlikely that intercohort differences would take similar forms across a panel of disparate cultures, the possibility still remains that the age differences mapped by our data have a generational rather than a developmental basis. That is, they may reflect generational, cultural differences *between* age cohorts, rather than life-cycle differences *within* individuals.

In order to sort out the contributions of nature versus nurture to these results, the TAT cards were readministered to the Druze and Navajo subjects (after a lapse of four years in the Navajo case, and after a lapse of five years in the Druze case). Surviving members of the original panel—those who had responded to our battery at $Time_1$—were sought out, reinterviewed, and exposed to the original TAT cards. The elicited stories were then analyzed and sorted into their appropriate mastery categories. These indexings were done independently for each story, without reference to the mastery status of the respondent's $Time_1$ story to the same card, and without reference to the subject's responses to other cards, at either $Time_1$ or $Time_2$. For each readministered card, it was hypothesized that the $Time_2$ data would show a pronounced shift along the mastery continuum in the direction of Passive and Magical Mastery.

When the $Time_2$ were arrayed against the $Time_1$ data for the Rope Climber card, the predicted shift was shown to have occurred, and at

FIGURE 3.2

The Rope Climber Card:
Comparison of Time₁ and Time₂ Responses
Given by the Same Navajo and Druze Subjects

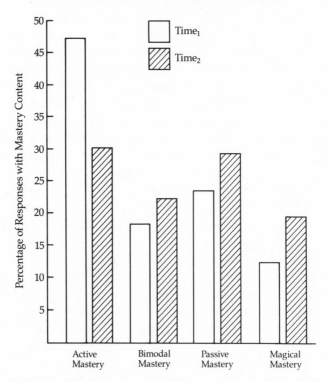

a highly significant level. As can be seen from figure 3.2, the longitudinal progression across the mastery spectrum found in the highlands of Golan and Galilee matches that found in the highlands of Arizona: In these very different cultures, the Time₂ mastery profile, when contrasted with the Time₁ profile, shows a pronounced and predicted shift to the right, in the direction of the more passive and magical orientations. In both cultural groups the basic longitudinal shift is along the same axis, from the internalized to the externalized experience of aggression. For both the Druze and the Navajo respondents, the most direct, competitive expressions of aggression recede over time, only to reappear, in later life, in a disguised, externalized form: the paranoid threats coded in Magical Mastery (MM₁).

A comparison of figures 3.1 and 3.2 provides further support of the developmental interpretation of the age differences found originally in the cross-sectional Rope Climber data. Note that the Time$_2$ mastery profile developed from the total, across-age longitudinal sample matches very closely the profile originally developed by the older (60+) cross-sectional sample. Besides matching each other, the two profiles are alike in their degree of discrepancy from the younger men's cross-sectional and longitudinal profiles. Both the older and the later profiles are low on the Active and Qualified Active Mastery positions, and high on the Passive Mastery (Constricted type) and Magical Mastery positions. Clearly then, the intercohort differences originally found in these data match each other. They are based on intraindividual shifts of a developmental nature, not on secular differences between generations.

The Heterosexual Conflict Card

The Rope Climber card distributions tell us that younger men are relatively comfortable with their own aggression, experiencing it as a resource, but that older men experience it as split off from themselves, as an alien threat rather than an internal resource. The age changes revealed by the Heterosexual Conflict card indicate that in later life women can become for men the embodiment of that external threat; they can present the aggressive visage that men have come to deny in themselves.

Let us consider in detail the responses to this card, which was developed as part of the standard Murray TAT battery, and which depicts a young man half-turned away from a young woman, who reaches toward him.

ACTIVE MASTERY. The stories grouped under this category show clear sex-role polarization: The young man is an energetic, assertive, and sometimes impulsive figure whose movement is centrifugal, "out of the card," away from domestic space, and away from the encircling, retentive arms of the young woman. He is moving toward adventure, challenge, combat, or discovery, beyond the reach of home and community. The young man is unconflicted or unambivalent about such activities and purposes; if there are doubts and fears concerning his enterprise or his aggression, these are located in the young woman rather than in the young man.

In some stories, the struggle between the opposed tendencies is more directly expressed: The young man asserts his strivings for autonomy through a full, open conflict with the young woman, and even a rejection of her; while she desires love, the young man, in pursuit of freedom, pushes her away.

QUALIFIED ACTIVE MASTERY. These stories are built around the same basic stem as that identified in the Active Mastery responses. However, in these cases the young man's aggression is either allied to suspect purposes—for example, "He wants to get drunk"—or it will lead to bad outcomes: "He could get hurt, or killed." The young woman plays an exclusively restraining role; her fears are considered to be justified, on moral or realistic grounds or both (QAM₁).

In the second subcategory under this heading, the emphasis is on an unspecified conflict between the young man and the young woman: He rejects her, but there is no explicit theme of a search for autonomy—the rejection of the woman is not in the service of some larger pursuit of freedom by the young man (QAM₂).

(Frequently in such cases, the aggressive theme is not spontaneously developed by the informant but emerges only in response to cues and promptings by the interviewer. In effect, the interviewer in these cases has taken some responsibility for the aggressive content tentatively expressed by the respondent.)

BIMODAL MASTERY. These stories contain clear and sometimes conflicting active and passive themes. For the most part, these are stories in which an active theme is first proposed, then left undeveloped, to be finally refuted by a contrasting passive theme that is not integrated into the prior text. Alternatively, while active tendencies may be integrated into the text, they are subordinated to an essentially passive plot and outcome. In many such instances, the theme of male aggression is retained, but its bases, along with the politics of the man-woman dyad, have shifted. The man's aggression is not intrinsic to him but is instead reactive to some assertion or rejection on the part of the woman: The young woman is ordering the young man around, she has refused to cook his dinner, or she has cheated on him with another man. Her aggression is prior to and explains the young man's aggression: He reacts to her initiatives rather than having her react to his. The man may be angry, even active, but his action does not serve any prior goal; having lost control over his goals and over the initiation of his action, he is essentially passive.

PASSIVE MASTERY (SENSUAL). In these stories, the theme of conflict between two sexually distinct figures has been replaced by themes having to do with mild enjoyment equally indulged by two undifferentiated figures: The young man and the young woman both look out toward a pleasing view, or they stroll together affectionately. There is no tension between the figures or between them and their surroundings.

PASSIVE MASTERY (CONSTRICTED). In some of these stories (PM_1) there is no mention of conflict between the young man and the young woman; aggression takes the form of an unspecified, threatening, outer-world agency that menaces the young people equally. The young couple's response to the threat is also undifferentiated by gender: Both "look at it," and neither of them takes any distinctive, effective action against the danger.

In other stories grouped here (PM_2), the young man is not aggressive but is instead a victim of impersonal aggression that has an internal source: He is sick, or drunk, and is accepting help from the young woman. The power relationships seen in Active Mastery stories are here completely reversed, with the young man weakened and the young woman as the stronger, protective figure.

Finally, in the PM_3 subgroup, there is no plot, merely a terse description of the postures and sex of the figures without any interpretation of their relationship or feelings: "He looks away from her; she looks toward him. I don't know what they are doing there."

MAGICAL MASTERY. In these stories, the aggressive, destructive implications of the card are either grossly accentuated or grossly denied, in both cases at the expense of realistic stimulus interpretation. In those instances (MM_1) where the sexual or aggressive possibilities, or both, are overemphasized, going beyond the realistic stimulus possibilities, the young man may be misperceived as an old man, a victim of time's "aggressions"; alternatively, he may be seen as a rapist or as an incestuous father who forces himself sexually upon a female victim.

For stories in the MM_2 subgroup, the sexual and aggressive card possibilities are utterly denied: The young couple may be seen as "angels," or the young woman as the Virgin Mary. In either case, the "sacred" imagery implicitly rules out any grossly sexual or combative possibilities.

As shown in figure 3.3, the distribution of story themes by age and mastery style demonstrate a clear age trend, one that is statistically highly significant. Across cultures, younger men—those in their for-

FIGURE 3.3

The Heterosexual Conflict Card: Distribution of Responses by Age and
Mastery Style (Kansas City, Navajo, Druze, and Maya)

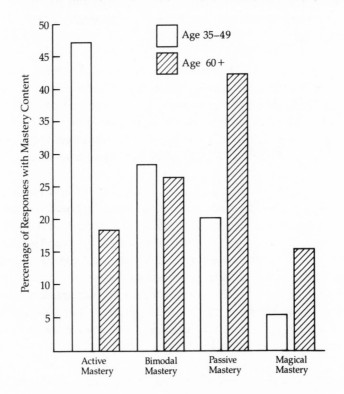

ties and early fifties—favor the Active Mastery and Qualified Active
Mastery themes; if they approach passivity, they do so tangentially,
through Bimodal Mastery constructions. They either stress the active
interpretations, or they appear to be trying out, via bimodal interpretations, their mingled active and passive responses to the latent
stimulus demand of this card. While there is a good deal of thematic
variation within the stories told by the younger group, in contrast to
the senior men they describe definite sex-role boundaries and an
outward-turned, aggressive, and centrifugal young man. In their view,
the young man avoids the nurturance offered by the woman and the
comfort offered by the domestic setting, in favor of aggressive action
on some extradomestic frontier. Young men are supposed to concentrate the aggressive potentials latent within the domestic zone, to
"export" that aggression away from the vulnerable home range, and

to discharge it against enemy, prey, or physical challenge, on some foreign ground.

Older men dance to a different music: In their interpretations of male-female relationships and tensions, the aggression suggested by the card is no longer exclusively packaged within the male but is either completely denied or distributed into his surroundings: It becomes a property of the young woman or of some looming, nameless threats located beyond the margin of the card.

As registered by the Heterosexual Conflict card, younger men see male aggression as exuberant and outward-directed, tending away from the domestic zone and toward extradomestic targets or challenges. But for older men, the male figure is more likely to be the victim of aggression than its explosive center. He is menaced by external threats, which he counters by retreat rather than advance: Thus he is harassed by his wife, or he is attacked internally, from within his own body, by illness or old age. The aggression that is viewed by younger men as an internal motor and center of the self becomes in later life an impersonal, external threat that acts to diminish the self. It may be a central fact in male aging that the wife comes to represent, as its outward metaphor, the aggression that has migrated away from the man. In subsequent chapters I will show that older women do indeed develop in their own right so as to justify the projection of male aggression toward them.

Longitudinal Studies

In order to rule out the possibility that the age and mastery theme distributions generated by the Heterosexual Conflict card were an artifact of cohort differences between generations rather than developmental staging within individuals, the card was one of those administered (after a four-year lapse in the case of the Navajo, and after a five-year lapse in the case of the Druze) to still-accessible Time$_1$ subjects.

As with the Rope Climber card, I predicted that the Time$_2$ distributions would show a decisive shift away from Active Mastery and toward Passive or Magical Mastery.

As figure 3.4 indicates, this expectation was borne out in both the Druze and the Navajo data, to a statistically significant degree. We seem to be dealing with orderly and predictable psychological change

FIGURE 3.4

The Heterosexual Conflict Card: Comparison of Time₁ and Time₂
Responses Given by the Same Navajo and Druze Subjects

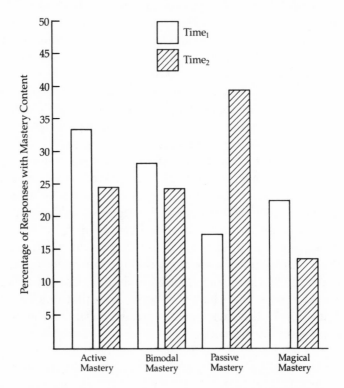

within individuals over time, not with intercohort cultural differences. Again, as in the case of the Rope Climber card, the longitudinal replication allows us to draw developmental inferences from the original, cross-sectional data.

The cross-sectional and longitudinal age-by-mastery-theme distributions of responses to the Heterosexual Conflict card extend the picture that we have already developed. In particular, the data from this card point up the increased subjectivity of later-life thinking, and shed more light on the later-life deployment, by gender, of "masculine" and "feminine" traits: As older men take on traits that were once exclusive to women, women take on, in the eyes of older men, some of the aggressive "masculinity" that these same men have relinquished.

In sum, while societies may vary in the degree to which they sponsor one or another mastery position, the passage through these stations in the latter half of life seems to be an individual and developmental matter, independent of culture. Across cultures, the processes of aging work to bring about a shift from an internal to an external locus of male aggressive power. In the course of aging, aggression is converted from an internal to an external experience, and the wish to be a center of power and influence is replaced by the wish for favor from powerful, potentially benign, but also potentially hurtful authorities. While this desire may be enacted in many different ways (depending upon personal and cultural circumstance), older men may come to feel, in their deepest awareness, that they are no longer centers of influence and control. Instead, these centers lie outside of themselves, consisting of powerful institutions and persons (including, as later chapters will indicate, their wives) who cannot be shoved aside or intimidated. Instead, their favor must be won, their dictates must be obeyed. Accommodation (Passive Mastery) replaces Active Mastery as the dominant male psychological style.

Longitudinal studies that make use of all TAT responses given by the Navajo sample further illustrate the major points made here concerning the individual's age progression through the stations of the mastery typology. Figures 3.5 through 3.8 graph the results of our procedures, in which all TAT responses given by the Navajo informants to all TAT cards are categorized by age, by mastery type, and by time of administration. The mastery profile generated by respondents who were in the age range 35–49 at $Time_1$ (fig. 3.5) shows that, as a group, they peaked on Active Mastery, and that there is little change from $Time_1$ to $Time_2$. The $Time_1$ profile predicts with exactitude to the $Time_2$ profile. But for those respondents who were in the age range 50–59 at $Time_1$ (fig. 3.6), there is a moderate but distinct swing to the right, toward the Passive and Magical Mastery stations. This movement is intensified in the most senior groups (figs. 3.7 and 3.8), so that for the 70-year-old cohort, the separation between the $Time_1$ and $Time_2$ profiles is almost complete. These findings from the longitudinal data again support our central contention: The age differences picked up in the cross-sectional data point to intraindividual rather than interindividual cohort differences in mastery orientation.

These findings from longitudinal data also point to a hitherto unpredicted phenomenon, one that could not have been picked up from cross-sectional data: Aging not only brings about a movement from active to passive; it also brings about an acceleration of this movement—the older the subject, the more rapidly he traverses the

FIGURE 3.5

Distribution of TAT Stories
by Mastery Orientations at Time₁ and Time₂ 35- to 49-year-old Navajo
(Time₁ Ages)

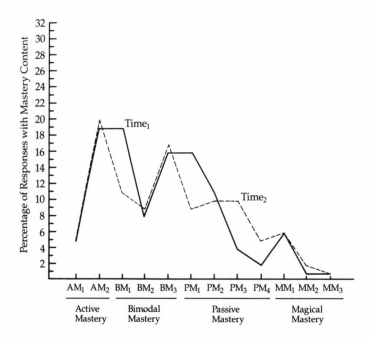

mastery track toward the passive and magical positions. The closer the lip of the waterfall, the faster the current; the closer we are to death, the more rapid is the sweep through the mastery positions. As figure 3.9 indicates, Navajo men who did not survive to give a Time₂ TAT protocol were more likely than their surviving age peers to have generated a Passive Mastery TAT profile at Time₁. The men who at Time₁ already stood at Passive Mastery (particularly the Constricted phase of Passive Mastery) were, in effect, closer to death than the Active Mastery survivors. Apparently, the depressed state captured by the Passive Mastery (Constricted) category is not only a metaphor of death, it is a prelude to actual death.

(The cross-cultural data from another TAT card that illustrates a special aspect of male aggression is considered in the appendix, under "The Horse and Man Card." The appendix also contains additional data about the Rope Climber card and data elicited by the Male Authority card, under the heading "Aggression-eliciting TAT Cards: Intercultural Analyses.")

FIGURE 3.6

Distribution of TAT Stories
by Mastery Orientations at Time₁ and Time₂ 50- to 59-year-old Navajo
(Time₁ Ages)

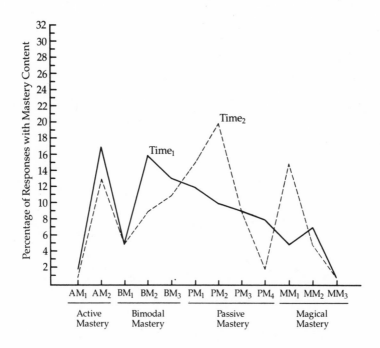

Depth Psychological Studies
by Other Investigators

There are few studies of aging men in other societies, or even in this one, that utilize projective techniques. Those that do exist tend to support the results reported here.

For example, Oscar Lewis (1951), in describing the results of Rorschach tests administered in Tepoztlan, a Mexican village, reported that men are "likely to be more exuberant than women, but also more anxious and insecure. As they grow older they lose their dominant position and the older adults appear disturbed, impulsive, and anxious. They seem to be losing the grip on society that the older women are taking over."

FIGURE 3.7
Distribution of TAT Stories
by Mastery Orientations at Time₁ and Time₂ 60- to 69-year-old Navajo
(Time₁ Ages)

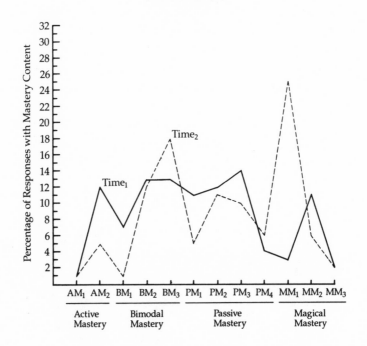

Leonard Giambra (1973), analyzing fantasy data from American populations, found that male daydreams regarding heroic action as well as personal advancement decline steadily with age, being highest for the 17–22 age cohort and lowest for the subjects in the age range 66–77.

Joel Shanan (1985) carried out an important interethnic, intergenerational comparison study of aging psychology in Jerusalem, his sample consisting of Sephardic and Ashkenazic Jews subdivided by sex and level of education. He utilized interview, questionnaire, and projective measures. His measures indicate that older men—by contrast to the younger, regardless of ethnic background or level of education—show a reduction in future orientation and in "active coping," as estimated by two independent instruments, the sentence completion test and the TAT. Jordan Jacobowitz (1984) has done longitudinal studies with Shanan's data and found that unconscious active

FIGURE 3.8

Distribution of TAT Stories
by Mastery Orientations at Time, and Time,
Navajo 70 Years Old and Older (Time, Ages)

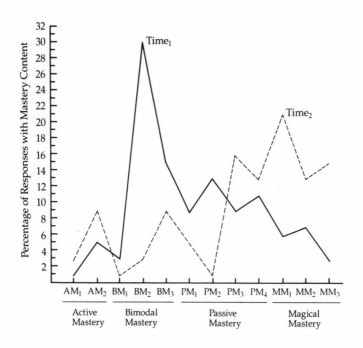

or passive orientations, picked up only by the TAT at Time,, became evident and pervasive in the more public, self-conscious interview materials at Time,, after a five-year lapse.[1]

Jacobowitz's findings are important on two counts. First, they give robust support to a central thesis of this work: that surgent unconscious contents, after staging through intermediate fantasy expressions, eventually find real, nonfantasy sponsors in the outer world, and move toward conventional forms of communication and behavioral expression. Second, they shed light on a nagging controversy in geropsychology between those who study aging personality through qualitative, usually projective means and those who stick to more "objective," self-reported, easily quantified instruments. The hard-nosed scientists are temperamentally uncomfortable with "soft" instruments, but they generally explain their objections on more rational grounds. They note that in studies which utilize both methods,

69

FIGURE 3.9

Distribution of Time, TAT Stories
from Navajo Deceased at Time₂ and from Navajo Still Living at Time₂

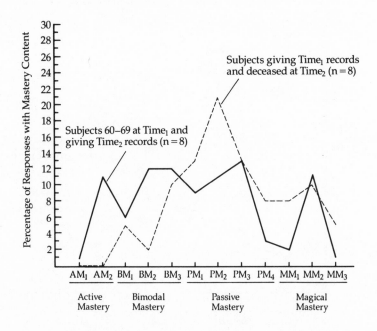

the findings from projective tests rarely predict the findings generated by "objective" instruments. Since inner fantasy does not match outward action, why bother with the soft, invalid projectives? As Robert McCrae and Paul Costa put it (1984), "Neither self-reports nor direct observations of behavior show parallels to the developmental sequence inferred by Gutmann from TAT responses." Faced with these discrepancies, Costa has regretfully announced that they sound the "death-knell" for any theory of substantial personality change in later years. What Costa and McCrae overlook is that fantasy is not a direct correlate of behavior, but is its antagonist within the personality system. As such, fantasy is part of the system of delaying functions on which the human ego is founded. As stated earlier, fantasy serves to delay and neutralize potentially dangerous impulses by keeping them out of direct behavior. Thus, Jacobowitz, who is more sophisticated about the nature of fantasy as an alternative to overt action, shows us that the correlations between fantasy and objective data do exist, but they are revealed only over time, after fantasy has exercised its proper function of reining in (and rehearsing) eruptive, potentially dangerous behavioral potentials. By the same token, the fantasied "death-knell" was also premature; perhaps it too should have been delayed.

Artistic Imagery

Thus far, I have reviewed the developmental shaping of the life cycle in the later years of the "ordinary" human, and the major data source in tracking such sequences has been the inner-life imagery drawn from our informants by projective instruments. Let us now consider individuals who do not merely report their fantasies but take a more active, reworking stance toward their inner life. These are the creative artists, who externalize and form their internal fantasies into visual experiences, not only for themselves but for all of us. Our assumption is that while artists may differ from the nonartistic majority in that they actively manage their inner life, these interior contents—their preoccupations, their appetites, their excitements—will be like those of "generic" men in the same life stage. The artist's creative act may bring a new product into being, unlike anything that has existed before, but despite the artist's unique technique and excitement, the work grows out of the same soil and is shaped by the same concerns that prompt the more conventional actions of nonartists.

The artist externalizes his fantasies to achieve a special developmental outcome, but his fund of imagery is similar to that of his nonartistic age peers who retain their fantasies internally. In the cases of both artists and nonartists, the fantasies give clear readings of their current psychological concerns and transitions. Thus we assume that a comparative review of works turned out by long-lived artists in their early and later years will demonstrate, through yet another medium, the psychological transitions that we have already identified in our cross-cultural samples. Moreover, we expect that artists will demonstrate these passages more vividly, that their works will provide clues to tendencies and transitions not observable in our usual data pools, and that they will show us the changing nature of the creative process itself, across the life span.[2]

The paintings and drawings used in the following analyses were found in representative collections of the various artists' works; and in the case of the cartoonists, in back issues of the *New Yorker* magazine. In all cases, the full sample of drawings available from a particular artist was reviewed in order to get a sense of his idiom, his preoccupations, and his range of content and technique. The individual works were then ordered into thematic categories whose selection criteria reflected distinctive characteristics of the artist's work as well as my own theoretical concerns. Thus the category system set up to accommodate the data from the cartoonist George Price reflects the prominence of vividly drawn women in his published works (particularly in the later cartoons) and the author's interest in the staging, by age, of male-female relations.

The investigation began with a study of two *New Yorker* cartoonists, George Price and William Steig, both chosen because they have had long and productive lives, because their superior draftsmanship merits the designation "artist," and because the approximate date of each drawing is easily determined from the *New Yorker* issue in which it appears.

William Steig: From Feisty Boys to Queen Mothers

The reader will remember that the cross-cultural data developed by the Heterosexual Conflict card portrayed two versions of the dominant woman, as she is conceived by men: We saw the "good mother," the man's haven in a sea of troubles, and the "bad mother," who dominates or disappoints the man, causing him, full of impotent anger, to leave home. The Steig materials trace, over time, the thinning out of aggression-laden, male-oriented themes, in favor of later-life depictions of the "good mother."

Reviewing the early work of William Steig (born in 1907), we find a group of cartoons featuring little guys who are trying to make it in the world of Real Men. In a locker room crowded with large, muscular males, a spindly little fellow tries to get their attention by loudly singing a World War I song, "Madamoiselle from Armentieres." Another cartoon series, entitled "Small Fry," sympathetically portrays boys who test their courage and mimic adult male behaviors. A later series, though still from Steig's early years (these produced during World War II), is titled "Dreams of Glory" and depicts—again, with a kind of amused sympathy—the grandiose dreams of small boys intent on impressing their big brothers. Thus a jubilant little cowboy, brandishing man-sized six-guns, is shown breaking into Hitler's headquarters to capture him single-handedly.

FIGURE 3.10

The Cartoons of William Steig: Age and the Depiction of
Male-Female Relations

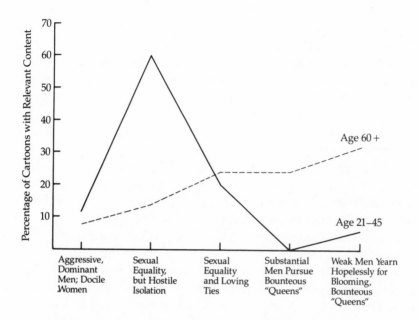

However, upon entering his middle years, Steig seems to turn against the aggressive masculine ideal (see fig. 3.10). He produces a series of drawings that ridicule armed, warlike men as posturing (albeit dangerous) fools. During the same early middle period, there appear a large number of drawings depicting angry men, alienated from a world they never made; men resolutely turned against women; men and women struggling together in a tight, entrapping net; and men and women as discordant, opposing parts of the same larger body.

The sequence of Steig's drawings, from young manhood to middle age, reflects a continuum formed around typical masculine concerns of the earlier life period: As a younger man, though doubting his own credentials, Steig tries to gain acceptance in the world of men from the big guys, the gatekeepers already established there. However, by early middle age, though aggression is still an important variable, Steig appears to have turned against his own assertive strivings and now disparages masculine qualities in cartoons of powerful but pretentious or stupidly destructive men. In his young manhood, Steig gently ridicules his own wish for acceptance by the big guys, but in his middle years he turns against those who have made it as men, and instead mocks the aggressive masculine ideal that he once emulated. In effect, his mid-life drawings reproduce the themes of Bimodal Mastery.

But even as he rejects the aggressive versions of masculinity, in his middle years Steig also seems to be troubled by an emerging femininity—represented by a man and a woman at odds within the same body—that he seems to recognize, uncomfortably, as part of himself. In other words, having put aside the idea of aggressive masculinity, Steig discovers—like his age mates around the planet, and with a kind of horror—a hidden quality of softness and femininity at the very core of self. In him, we see the conflicts over emerging androgyny that are generic to this midlife period.

However, by the sunset years, the crisis of androgyny appears to be over. In his sixties and early seventies, Steig's drawings show a profound change in content and execution. Thus his line, which was once definite, aggressively boundary-defining and cutting, is now open, rounded, and impressionistic. The shift in his later style, toward a more soft and sensual line, is very much in keeping with the drastic content changes of the same period toward the notable idealization of bounteous, queenlike women—Venuses attended by diminished men. Whereas in the early years no more than 10 percent of Steig's cartoons depict easy affiliation between men and women, in his elderhood at least 35 percent of the sample portray comfortable male-female relations, and almost half of the drawings of this period feature radiantly sensual and buxom women, either alone or attended by (usually) adoring men. The aging Steig, like the generic men we have been studying, divests himself of aggression, or uses it to obliterate the masculine rather than the feminine principle. Older now, he tries to please the mothers rather than the fathers. When he was more interested in masculinity, the younger Steig wanted the affirmation of men; but when the older Steig has come to terms with (and even relishes) his more feminine side, his wistful little men try to link themselves to a "good mother," but one who forever eludes them.

George Price: From Active Men to Diminished Men

George Price (born in 1901) traces a similar artistic continuum across his productive life: from the drawings of "masculine" subjects in the early years, to those of the later years that feature powerful women rather than the relations among competing men (see fig. 3.11).

FIGURE 3.11

The Cartoons of George Price: Age and the Depiction of
Male-Female Relations

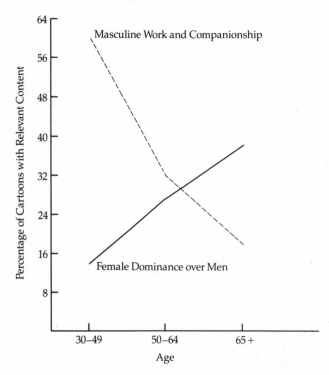

Thus, when Price was between the ages 30 and 49, 60 percent of his cartoons had to do with the World of Men: males in competition or in collaboration, mainly in outdoor, public space, with women either absent or relatively insignificant. A typical cartoon of Price's earlier years shows a group of straining runners at the moment of breaking the finish-line tape, while a man in the background clocks them, but with an hourglass. In another cartoon from this period, two men holding strike signs picket a surgical supply store, in wheelchairs, while the angry owner looks on. Despite their overt differences, both cartoons have in common a struggle among men, observed by men, in outdoor, public spaces that are essentially devoid of women.

But when Price reaches his early fifties, striking shifts take place. The majority of his cartoons no longer depict the World of Men, and women are more often dominant in the drawings that depict inter-sexual relations. Consistent with this shift toward empowered females, Price also shifts, in the depiction of space, toward the now preferred domestic ecology: Two-thirds of the later drawings depict indoor and household rather than outdoor scenes.

A typical cartoon from Price's middle years shows three generations of men—a son, a father, and a grandfather—frantically pedaling their bicycles around a living room table, while the woman of the house, joined by a female friend, looks on dubiously. She remarks: "It's not enough I let them have a six-day race. Now they want I should put up money for sprints!" Another cartoon from this middle period depicts an elderly man in cowboy costume sitting on a spavined horse, shading his eyes as though gazing out over vast ranges. In actuality, his horse stands within a small, junk-filled, fenced-in backyard that also contains a female visitor to whom his disgusted-looking wife complains, "You'd think it was a ten-thousand-acre estate he's looking over!"

These are, in effect, transitional cartoons. In both cases, men act in ways—competitive (bike-riding) or ascendant (horse-riding)—that are appropriate to the male arena but not to the domestic zone, where they come under the disparaging eyes of the resident women. In both cases, the humor has to do with the fact that men are trying, inappropriately, to establish the "heroic" style within ecologies that do not provide scope or appreciation for this style and that underscore the pretentions of the now comical male.

Later, in Price's seventh and eighth decades, the sex-role and milieu reversals are almost complete. The World of Men becomes a relatively neglected arena, four out of every five cartoons are concerned with male-female relations, and the female is dominant in at least half of these, as well as in the plurality of all drawings from this age period. Again, consistent with the shifts in Price's depiction of intersexual relations, the "ego space," the arena within which the artist visualizes action, becomes exclusively domestic: one room, without doors or windows to suggest an outdoor world, and this space crammed with the presence of a hostile older wife. In a recent cartoon (fig. 3.12), the mild husband and the burly wife wear matching T-shirts. Hers reads "Fight," while his reads "Flight." The elements of the male emergency response—to attack or to flee when the battle is lost—have been parceled out among the sexes, but the usual allocation by gender has been reversed.

Other cartoons of Price's late period are of this same order. A harassed older man, sitting in an armchair with a can of beer, stares fixedly at his TV set. His shrewish wife, arms folded, stands above

76

FIGURE 3.12

Drawing by Geo. Price; © 1984 The New Yorker Magazine, Inc.

him and snarls: "You say you're sorry. You act sorry. And you *look* sorry. But you're not sorry." Similarly, a mild-looking little man sits at a kitchen table forking food, while placating his stony-faced wife: "You are right, honey. It *is* too good for the likes of me." In both cases, the men are ensconced within domestic space but disadvantaged there, themselves inoffensive but oppressed by their hostile wives. Save for the TV set as a kind of window, there is no indication of an alternate, more public, or more "masculine" region. In one case the man adjusts to his condition through accommodation and submission: He grants the correctness of his wife's complaint against him. In the second case the man opposes the wife, but passively; using the TV set and the can of beer, he makes an inward migration out of the scene. In effect, he escapes the wife by focusing on the impersonal gratuities that come to him through the eyes and mouth.

As younger cartoonists, Price and Steig made humor out of the predicament of younger males: their struggles with other men. But in their later years, they make humor out of the universal predicament

77

of men among tough females, the women who have replaced men as the powerful figures in the older man's world. Price and Steig have externalized, into their respective renditions, the generic fantasies of older men, and these portray personal vitality in an alienated form: as a property of women rather than men.

Between them, these artists show us the two images into which aging men, universally, have split the energized maternal image: Steig shows us the bounteous but essentially unobtainable woman, the focus and wellhead of good power, while Price shows us the dangerous "witch," she who concentrates bad power and who replaces other men as the competitor, even the enemy, of the aging male. Unmindful that it is ultimately men who grant women their dangerous, mythic power, Price attempts, through his cartoons, to make older men (and himself) feel better about their subjugation. He turns the powerful woman into the hag, the "battle ax." She is powerful, he says, but she is also stupid; by fooling her or by infuriating her, one can win the victories of the weak. In this war, Price seems to say, men can gain some version of victory by creating their own supplies, the belly pleasures and eye pleasures of booze and boob tube, thus denying their need for the wife/mother. For example, a notable recent Price cartoon shows a happy older man walking into his den, clutching a long flagon of beer as he looks expectantly toward his snacks and his video game. There is no woman in sight, and he wears a T-shirt captioned "Living Well is the Best Revenge."

Edward Hopper: From Outside to Inside

The age decline in portrayals of male aggression is not limited to an artistic subspecialty or a unique historic cohort of artists. Thus, in our time, longevous artists who are in no sense cartoonists show the same progression as Price and Steig: from being centered on competitive and productive "masculine" concerns, to being centered on women's works and on the fate of men in the interior, domestic world of women. Consider, for example, the life work of Edward Hopper (1882–1967), a contemporary master of some importance, who painted continuously—again, without any loss of his powers—from his youth until his death in the ninth decade.

Figure 3.13 indicates that, as a young man, Hopper painted, using his own artistic idiom, the World of Men: impersonal action by large, powerful machines (locomotives, tugboats, etc.). However, in his middle years Hopper developed a tentative interest in the domestic

FIGURE 3.13

The Art of Edward Hopper: Age and the Settings (and Gender) of Action

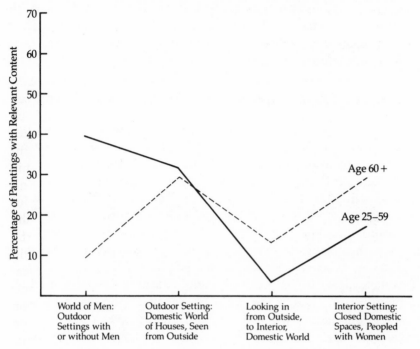

world—the world of houses—but these domiciles are rendered from the architect's perspective, as part of the technical, impersonal world. One of Hopper's most famous middle-period paintings is of an imposing, white Victorian house that stands alone, isolate in its own special air. In the foreground, between this piece of encapsulated domesticity and the observer, there runs a railroad track, yet another symbol of the technological order. In this instance, Hopper is clearly drawn to a domestic structure, but he keeps a safe distance from it, placing between himself and it a token of the active male world. But as Hopper gets older, he gets bolder: He approaches houses more closely, so that their outer walls fill his canvas. At this point, although still an outsider looking in, Hopper is intrigued by events in the domestic interior rather than by the exterior architecture. Now he is looking in through the windows from the outside, to discover and paint a warm, interior world inhabited by women. However, he still keeps a buffering pane of glass between himself and such exciting yet troubling sights.

But in his last painterly years, Hopper (like Price) has completed his journey from the vital outside to the quiescent, "womanly" interior: In the last paintings, he is inside, looking out; he is fully within the house, ensconced in the domestic zone. His ego now lives in an enclosed space, one typically furnished, at its very center, with a compelling (often nude) woman.

Jean-Auguste-Dominique Ingres: From Heroes to Harems

A cross-sequential comparison shows that the life-span journey of the artist, away from aggressive "masculine" motives and settings, is not limited to the current generation of cartoonists and painters. The observed changes in artistic emphasis take place within the span of individual life histories rather than across the span of collective history, of generations. Consider the life work of Jean-Auguste-Dominique Ingres (1780–1867), a French master of the early-nineteenth-century classicist period. As we would predict, the young Ingres painted land battles, sea battles, and the deeds of mythic heroes. If women appear in his youthful works, it is only as victims of male-inspired warfare and as the occasions for manly heroism: Helpless, they can only adore while the gallant knight saves them from the dragon. But again, over the life span, time brings about a reversal of gender priorities (see fig. 3.14). As Ingres ages, his males become weakened, inwardly troubled. They may have public stature, but they do not seem to have an inner fortitude to match it, and women are depicted as the true sponsors and instigators of male accomplishment. Finally, true to form, in the seventh and eighth decades, Ingres paints, almost exclusively, the virtues and worlds of women: nursing madonnas and the magnificent odalisque canvases—the signature paintings of his entire career—which limn the sealed, inner world of the harem, stuffed with lovingly painted, opulent female flesh. We see the typical development, over the life span, of the ego space, as Ingres the artist externalizes it in his canvas. As a young painter, he calls up the outdoor, action-filled vistas of the male perimeter, and in his later years, he recreates the enclosed, hothouse climate of the emotional, woman-filled domestic interior.

In their rendition of space across the life span, these four artists have traced out a clear human universal. Whether cartoonist or fine artist, whether French or American, whether of this century or another, as male artists age they migrate in their creations away from

FIGURE 3.14

The Art of Jean-Auguste-Dominique Ingres: Age and the Settings
(and Gender) of Action

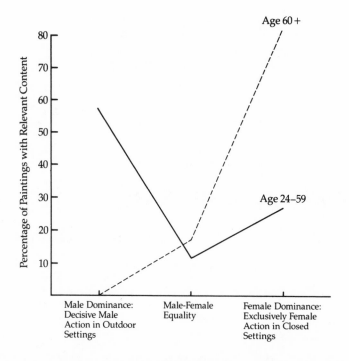

boundless outer space, away from the domain of the perimeter and large, thrusting, exploratory movement; instead, they carve out their claim to the interior. The projective tests allow older respondents to externalize surgent, unconscious concerns in their verbal interpretation of ambiguous stimuli, and the work of artists from different countries and eras shows how the same surgent potentials can move toward open, shaped expression to become part of the conscious experience of the artist and the aesthetic experience of his audience. In the next chapter, I will consider the social expressions, in age-graded roles and usages, that these masculine potentials can take, under various cultural conditions.

4

FROM WARRIORS TO PEACE CHIEFS: AGE AND THE SOCIAL REGULATION OF MALE AGGRESSION

If I were fierce and bald and short of breath,
  I'd live with scarlet Majors at the Base,
And speed glum heroes up the line to death.
  You'd see me with my puffy petulant face,
Guzzling and gulping in the best hotel,
  Reading the Roll of Honor. "Poor young chap,"
I'd say—"I used to know his father well.
  Yes, we've lost heavily in this last scrap."
And when the war is done and youth stone dead,
  I'd toddle safely home and die—in bed.
—SIEGFRIED SASSOON, 1886–1967*

Thus far, our findings are based on two kinds of fantasy: fantasy with aggressive content and fantasy with affiliative and peaceful content. All fantasy, whether aggressive or passive, is used by the psyche either as a substitute for action or as a rehearsal for future action. Again, the linkages between fantasy and action are complex and devious. Fantasies refer to surgent but repressed (because possibly dangerous) action tendencies that may never, except in pathological states,

* "Base Details" (1918), from *Collected Poems* by Siegfried Sassoon. Copyright © 1948 by Siegfried Sassoon. Reprinted by permission of Viking Penguin, Inc.

find their way into sculpted behavior. Or they may express action tendencies that will be revealed in future behavior, after all competing possibilities have been mentally rehearsed, after tolerant sponsors have appeared, or after the conditions, internal or external, that render the action dangerous have receded.[1]

The TAT results discussed in the last chapter reveal, over a variety of societies, a clear falling off in aggressive imagery as men age. The surgent aggression that is registered in such fantasies is most likely to enter public behavior in the most unambiguous form within warlike and predatory societies—the communities that honor successful male aggression. Accordingly, if we intend to trace the life course of aggression in direct action, that is, in other than fantasy expressions, we should first turn our attention to the ethnographic data from warrior and hunting societies, and review their age norms for aggressive behavior.

Universal Themes in Age Grade Systems

Age norms come about when conventional expectations about proper behavior are applied differentially to generational groups; the resulting sets of codes form age-grade systems. As Adriaan Prins (1953) puts it, the age grade exists as a formal frame defining behaviors, rights, and obligations into which the men of an age set move at the proper life stage. While these systems can be described as though they exist independently of those who man them, yet they are human inventions; they are not institutions created by some supraindividual "social reality" to guarantee its own continuity. They represent arrangements of the social life cycle crafted by those with sufficient institutional power to arrange human affairs according to their need, their convenience, and their view of the social good. In the small, folk-traditional society, the cadres that possess this enabling power, though composed mainly of older men, ordain the role expectations for all age grades, including those of younger men. In effect, the traditional older man brings about a major compounding of human nature and social nurture: The elders' views as to the right and proper staging of the life cycle—views based on their own urgings, fears, and ideals—are formally detached from their personal origins to become the tablets of the law, the structural armature of the society.

Although the reigning elders may claim to speak in God's name rather than their own, the ostensibly formal and objective age-grade system is nevertheless saturated with their own special priorities. If there is some standard, universal quality in the older man's relationship to aggression—his own and that of others—these orientations should find their expression in equally standard age-graded regulations, particularly as these bear on the management of aggression, and particularly in warlike societies—those in which the human capacity for aggression has been culturally sponsored and raised to a high pitch.

I will begin by contrasting and comparing the role expectations for young men with those for older men over a wide range of societies that are similar in their warlike nature and traditional characters, but different in their ethnic, demographic, and linguistic characteristics. My prediction is that the age-graded norms for young men's behavior will encourage one or another aspect of Active Mastery—the forceful subjugation of nature, or other men, or both—while the age-graded norms governing older men's behavior will express more pacific, Passive Mastery themes: accommodation to nature, to other men, to the gods, or to all of these.[2]

Young Men as Killers, Older Men as Peacemakers

The most detailed and sensitive study of age grading in a prototypical warrior society was carried out by the anthropologist Paul Spencer (1965) among the Samburu of East Africa (an offshoot tribe of the bellicose Masai). These people only recently gave up active warmaking, and the Samburu allegiance to the warrior tradition is reflected in the fact that they still maintain the age grade of unmarried warriors, the *Moran* (though these no longer do battle with other tribes). Samburu society is strictly age-graded, and young men enter their lifelong age set, the society of peers, at the time of their group circumcision, moving as age classes through the successive age grades that define the duties and prerogatives of manhood for all major life stages. The

Moran are bachelors, ostentatious in dress, erratic in behavior, highly conscious of honor and prestige, prone to having affairs with married women, and overly perfectionist in their demands on themselves and others. They could be Tybalt's troop, swaggering, lechering, and quarreling through the streets of Verona.

In order to check the energy and the predation of these young males, the rules of the society, as enforced by older men, confine them to special clubhouses off in the bush. The tribe is gerontocratic, respect for the older men is a pivotal value, and the age class of elders, who cannot match the *Moran* in physical strength or appetite for battle, enforce the physical limits that they have set with moral strictures backed up by supernatural power. The system of respect for the aged (*Mkanyit*) places the aggressive young *Moran* under the control of the elders; these harangue against the younger men in public meetings, reproving their excesses, and threatening them with the elder's potent curse. If the *Moran* openly challenge the elders, they must face the sorcery that the older men can level against them; by the same token, the elders admire, envy, and fear the fierce and decorated *Moran*, who have the power to seduce their wives. Thus the *Moran* fear sorcery managed by older men, and the *Mkanyit* fear adultery instigated by young men. The generational tension, here channeled between the age grades, does not phase out for particular age sets until the older *Moran*, in their thirties, become less assertive, more respectful to the aged, and ready to marry.

Spencer notes that this is a stable and recurrent African pattern. Elderly gerontocrats, practicing and perhaps enjoying polygyny, naturally provoke the rage of the sexually frustrated young men and then fence them off from the community, turning their warrior's rage outward against the communities of others. He finds the essentials of this design among the Rendille (neighbors to the Samburu), the Zulu, the Nyakyusa, and the Lele of Central Africa.

Thus, while the old men of these tribes might recognize that the intemperate young man is the best vessel for the collective aggression of the group, they also recognize that this aggression must be physically removed from the vulnerable precincts of the intimate community. If the domestic order is to be preserved, the aggression of younger men must either be hedged about with psychological inhibitions or fenced off behind physical barriers set between the assertive young males and the vulnerable members of the society. Among the Samburu, the defenses against dangerous aggression are externalized into social structures such as the age-grade system, rather than being

85

internalized in psychic inhibitions, and are distributed between rather than located within individuals. Given such assurance of strong external controls, the young *Moran* are freed to be outrageous; the task of confining their aggression is left not to themselves but to the elder age grade. Thus, in a society where the qualities of aggression and control are parceled out according to the age-grade system, aggression gravitates to the younger men, and cautious control, bolstered by moralism, is conceded to the older men.[3] The pressures of Samburu society may require this polarization, but species imperatives have the final say: They dictate how the aggressive and control functions are actually distributed by age and sex, not only for the Samburu, but for all warrior societies.

Prins (1953), who also studied the age structure of East African societies, has found parallels to the Samburu pattern in three tribal cultures: the Kikuyu, the Galla, and the Kipsigis. Among the Kikuyu (who recently, as the dreaded Mau Mau, terrorized white settlers), men in their middle thirties are in the grade of senior, advisory warriors; after passing their mid-forties they become members of the elders' council of peace. They end their days in the final grade, that of the priests. Thus the Kikuyu cut the same distinction between the extravagantly aggressive bachelor warriors and the more sedate, "bourgeois," householding elders that we found among the Samburu. Younger men generate their own power and thrust it outward, against human or other natural adversaries, while the older men deal with the threats posed to the community by supernatural power. The elders' task is to deal with breaches of taboo and the disastrous consequences thereof; in exchange for this sacred work, they alone are permitted to eat the sacrificial foods. As among the Samburu, the young men forge the rage that was fostered by their restrictive community into a weapon to protect the community and to menace the communities of others. The young men must learn to harness their own strength, while older men must, through ritual means, harness the strength of totemic sponsors, beat back challenges from their sons, and ward off malign supernatural power. While both age grades must deal with rage, the old men deal with aggressive power that has its origin in sources external to them, whether demonic, divine, or filial.

The same age career, from unchecked, exuberant aggression to sober restraint, is found by Prins among the Galla. There the youngest grade is *Daballé*, peopled by the age class of small boys. The *Follé* grade is occupied by the age class of boys who wear women's dress as a sign that they are not yet men. These prove themselves through quests and tourneys, earning the right to woo a woman. They also gain the right to use obscene language toward women, and they are expected to violate, without indemnity, the tribal codes. Thus encouraged, their aggression crests, to break over prey and enemy in the *Qomdala* grade. This is the grade of "young men"—the *real* warriors—who must conduct raids and learn to kill. They are not entitled to dance publicly with a woman and to marry. From there they pass into the grade of *Luba*, mature men who represent tribal authority, and they end their days in the ranks of the *Yuba*, the grade of older, cautious men, counseling the *Luba* against recklessness and passing on their life experience to others.

Consider briefly the ways of a third African group, the Nsukka Ibo of Nigeria. They celebrate many forms of accomplishment: not only skill in war but also prudence in human relationships. As Austin Shelton (1965) describes it, this is a dual-valued society, which legitimizes the high achievement and rivalry of youth, as well as the cautious restraint and social sensitivity of the aged. Each season of the life cycle finds its matching social values, in such a way that men, as they age, transit from one value consensus to another. The older man reaches a stage of life in which he leaves farming and manual labor to the young, on the grounds that he must now devote his energies to more important matters—maintaining order and justice in the clan, keeping his people under the protection of God and the ancestors, and teaching the young the correct ways of human relationships. The aging Ibo no longer stands as a representative of his own career and productive energies; instead he represents the beneficent power of gods and ancestors. This outcome seems to have good consequences for the mental health of the older Ibo: Shelton finds a very low incidence of what he terms "psycho-senility" among them.

But are we describing an African regional phenomenon, a consensus among African warrior tribes that assigns the craft of war to young men and the moral regulation of society to the older men? We may be citing a parochial feature of African age-grade systems rather than some human universal having to do with the developmental staging of aggression across the adult male life span. Siegfried Nadel (1952) shows us graphically what happens when African age-grade systems do not conform to some more universal age distribution, by developmental age, of aggression and control. Nadel contrasts the Korongo and the Mesakin, neighboring tribes of the northern Sudan. These societies are structurally and culturally similar: They are age-set societies whose young men, enrolled in fighting and warrior societies, live apart from the larger community, in the bush. In both the Korongo and Mesakin societies, there is a general male fear of growing old, a fate that is equated with loss of virility and manhood. In both societies, aging is explicitly seen as a process of feminization, the inevitable result of sexual intercourse. But there the similarity ends. The Korongo are without witchcraft beliefs, while the Mesakin are obsessed by them, their social life poisoned with suspicion and fear. Nadel argues that this crucial difference in community atmosphere pivots on specific but equally crucial differences in the timing of otherwise similar age-grade prescriptions in the two societies. A Korongo man does not enter the age grade of relatively quiescent elderhood until the onset of physical debility, at around age 50, and the social terminus of machismo, in the sixth age grade, coincides with its "natural" ending, in the sixth decade of life. On the other hand, Mesakin age-graded arrangements arbitrarily call for a snuffing out of aggressive masculinity at the end of the second age grade, so that a man becomes socially "old," is rendered officially passive, while he is still physically young, in his midtwenties. But masculine rivalry cannot be legislated away by social fiat. Nadel understands Mesakin witchcraft as a symptom of the continuing hostility between the resentful "elders" of the third grade and their nephews (sisters' sons) in the second grade. Because the two age grades must not resort to physical combat, they express their deep and continuing young man's rivalry via witchcraft accusations.[4]

The Mesakin case shows us that social nurture is not enough; age-graded norms designed to regulate male aggression in a particular society will promote harmony only if they incorporate the norms set by a common human nature that is present in some recognizable form in all societies. Thus, having identified the "African" arrangement, by age grade, of warrior societies, Spencer finds much the same writ in warlike peoples around the world. Specifically, he cites the Groot Eylandt of Australia, while Kardiner and Linton (1945) describe a pattern among the Comanche cavalry of the American Southwest that likewise echoes the African arrangements. The young Comanche braves initiated war parties, continuing in this role until their midthirties, when there was usually a falling off in their appetite for raiding. Two roads then opened up to the slackening warrior: He could become either the "peace chief" or the "bad old man" who substituted sorcery for the lost physical prowess of the warrior. The peace chief restrained the warlike ardor of the young men, reminding them of the casualties, the widows, and the grief that are caused by war; the bad old man, though he had passed the warrior stage, still kept title to his rage and rivalry, relying on sorcery rather than physical attack against the young fighters of his tribe. In this case, as in Africa, the older Comanche had the task of controlling external power. He either restrained the warriors or matched their innate physical strength with power borrowed from supernatural, totemic sponsors.

As among the Samburu, intergenerational male rivalry often continues, but the bad old men, still driven by that rivalry, move to a stance that combines Active and Passive Mastery. They manage the bimodality of older men, the potential conflict between active and passive modes, through defensive splitting: While conserving their masculine rivalry toward young men, they passively propitiate the gods. An increasing reliance on primitive thinking makes this solution workable. Through reliance on Magical Mastery, they project their own destructive aggression and experience it outside of themselves, in an alienated form, as a property of the gods. Then, in the passive stance, they propitiate the gods, humbling themselves before them, so as to be granted an endowment of their *tabu* power.[5]

In essence, the old Comanche enacted the process of projective identification, in which self-esteem is gained through illusory merger with the agent that they themselves have, in fantasy, empowered. In later life, identification with outside power sources seems to replace the younger man's active, even predatory attack on prey or enemy— for them, the external vessels of power.

Peasants: Young Competitors and Restraining Elders

Just as the concentration of aggression in young men is not limited to Africa, neither is it limited to warrior societies. Thus the conversations of Gisela Steed (1957) with the elders of a peaceful Gujurat (Hindu) village elicited views of the Indian life cycle that sound familiar. Even in this bucolic setting, younger men are expected to be angry, impetuous, and driven by a need to succeed; there follows a shift to a more peaceful nature in maturity, and finally a return of peevish but by now ineffectual anger and egocentricity in old age. This sequence is backed up by formal, explicit prescriptions of the

Hindu religion, which holds that young men should take up the vigorous, productive stance toward nature and other men, and that old men should cultivate their own spirit and their relations to God.

These phases are clearly defined, and passage through them at the proper time and sequence is a matter not only of convention but of religious observance. Thus the Hindu *Asrama*, or prescription for the life cycle, defines four major life stations: the student stage, the married householder stage, the "forest" or holy hermit stage, and finally the *Sanyasa* stage, involving the complete turning away from worldly attachments, toward the ethereal. In actuality, most men do not proceed beyond the householder stage, but their fixation at this level is excused, on the grounds that the active, managerial householder is the supporter of all other life stages.

From another Asian sector, Colleen Rustom (1961) reports that four major life stages, similar in nature to those named in the *Asrama*, are recognized by Burmese villagers. Until puberty, the task of the growing individual is to acquire the wisdom of respected parents and elders. The following, or "virgin," age is centered around the search for a spouse. This accomplished, the young husband's duty in the first stage is to acquire property for himself and to "give, show, teach" for others. Parents are expected to remain economically active until all their children have been married. In the postparental fourth stage, the older Burman leaves the mundane, productive tasks to younger family members and devotes himself to merit-acquiring religious deeds and to meditation. His inner life is expected to change in step with his altered public behavior: Presumably, the older Burman loses interest in worldly affairs, curbs his former appetites, and discovers an interest in Buddhist philosophy. This last period is also the peak of the life cycle. A long life is taken as proof of "good karma" and augurs well for the individual's next existence. The elders have provided for their children, and it is now the duty (as well as the pleasure) of the children to care for them. Not surprisingly, Rustom finds few Burmese old age homes, and these have small populations.

The same shift toward a worshipful old age is found in other Oriental village societies. For example, Donald Cowgill (1968) notes that, despite a continuing deference to the aged throughout their lives, Thai men over 60 are expected to pass on the active leadership of their communities to younger men. In the Thai villages, the real political leadership is in the hands of the middle-aged. Following their abdication of temporal affairs, the majority of older men will either take up or return to the Buddhist priesthood.

China conforms to the general Oriental (and global) pattern: Margaret Mead (1967), reporting on the typical life course in traditional China, tells us that the young men were expected to be active, enterprising, and ruthless. After such a sweaty, effortful adulthood, they were entitled to retire into a relaxed, meditative, and "noble" old age. In effect, young men in traditional China were expected to be as ruthless in civil pursuits as young Africans and Comanche were expected to be in their warfaring pursuits. Elaborating on the fate of the Chinese aged, Pow-Meng Yap (1962) reports that older Chinese men and women are expected to keep a vegetable diet, to abstain from sex, and to practice self-cultivation, worship, and meditation. In exchange for this "moral work," they receive food and shelter.[6]

Thus, in their versions of the life cycle, the peoples of the Far East cluster toward a consensus: Young men are expected to live in the mundane and pragmatic world, there to push and sweat energetically for success, but in later life men are expected to withdraw their attention from this arena, to restrict their action in this world, and to refocus their attention and concern to the next world and their relationship with God. In all these cases, the young man is the center of action and the older man is the center of contemplation, dwelling not on his own needs and deeds but on the action and the power of God, on powers and purposes more inclusive than his own. (The difference between warlike and pacific peoples seems to be that former warriors turn to the gods to renew their own strength, while aging peasants use prayer to renew the gods and the natural cycles that depend on them.)

This idea, that younger men should live fully within the productive or warrior style of their culture, while old men should concentrate on the beneficent activities, the "productivity," of God, is not a regional phenomenon parochial to East Asia. For example, I have found an equivalent program for the male life cycle among warrior-peasants of the Middle East, the Golan and Galilean Druze. The younger Druze men, whether farmers or soldiers, are industrious and enterprising, generally aiming to extend their holdings in land, orchards, and herds. They seek success for themselves in agriculture and advancement for their sons through education. The middle-aged peasant transfers many of his blunted personal ambitions to his son, and he undertakes significant sacrifices to further his son's education, in the clear hope that the son, as an educated man, will achieve the success and status that was denied the father. But this investment in a career, whether for self or for the son, dissipates after the father, having given evidence of an exemplary life, is invited to join the ranks of the *aqil*, the sub-society of religious elders. After his initiation into the religious society and after receiving his personal copy of the secret and sacred text of the Druze religion, the outward behavior as well as the inner life of the older Druze changes radically. He is less concerned now with the assertion of self and more devoted to the glorification of Allah. In this service, as noted earlier, he shaves his head, adopts special garb, gives up tobacco and beverage alcohol, devotes much time to prayer, and dwells on God's mercy while disclaiming his own past grossness, frantic appetites, and stupidity. It is now the power of God, rather than his own, that he idealizes, deflecting it, through prayer, into socially and personally productive channels.

The foregoing brief ethnography of formal age-grade systems suggests that very diverse cultures fashion their age-grade programs to accommodate the competitive, productive potential of younger men and the seasoned capacity for ritual or meditation of older men. In each society, the age-graded expectations for appropriate behavior seem to spring out of the social fabric itself, setting norms to which all proper men must conform. Nevertheless, the thematic solidarity that we find across these various sets of cultural rules suggests that age-grade systems are not independent arbiters of age-specific attitudes and behaviors; they are only the final, socially mediated expression of surgent energies that are shaped and initiated by species-level agents, themselves prior to and independent of culture.

The Age Staging of Productivity Norms

The behavioral manifestations of Active and Passive Mastery are not limited to formally age-graded societies. The same trends show up in those largely Western, or Westernized, societies that do not maintain traditional, explicit systems of age grades. In such settings, professions serve as the contexts in which we can observe the age-qualified progression through conventional social formats. Each profession, as a quasi-traditional subculture, maintains a range of norms to regulate proper professional practice and to limit the raw expression of competitive drives. And, as in the traditional folk society, we can observe the assortment, by age, of such rulings, with young men being the most responsive to the protocol for vigorous, assertive professional performance, and older men more apt to "hear" the implicit rules concerning the ethical and human aspects of performance, even when these are in opposition to competitive and profit motives. In non-age-graded societies, active-passive changes in personality may not be enacted through changes across formal roles (as from war chief to peace chief), but in modal changes within formal roles. The aging practitioner keeps the role title and responsibilities, while changing, often in subtle ways, the style, goals, and rewards of practice.

Irving Webber, David Coombs, and James Hollingsworth (1974) studied age differences in value orientation among the political leaders of three Colombian cities. Though the sites ranged from very traditional Popoyan to progressive Medellin, the age variation within cities remained constant: By contrast with the younger *politicos*, the older men (60 and over) were more interested in being than doing, in harmony with nature rather than subjugation of nature, and in present rather than future time. They had moved from a coercive, attacking stance toward one in which the leader appreciates the texture of events without trying very hard to change them.[7]

Likewise, Jay Abarbanel (1971), who studied generational patterns in the rural sector of Israel, found that younger farmers were much more achievement-oriented than their fathers. Their striving led to constant father-son friction over the proper management of the farm (the *moshav*). Presumably, this conflict reflects an age rather than a cohort effect. The now-cautious fathers were once the pioneering generation of Israel, the same socialist warrior-peasants who once battled successfully against the desert, the Arab armies, and the British empire.

Yonina Talmon-Garber (1962), a sociologist of the *kibbutz*, also noted the later-life growth in domestic, hearth-centered interests, although among the pioneer *kibbutzniks* this change makes for psychic trouble rather than tranquility. Studying the course of aging in the youth-centered *kubbutzim* of Israel, Talmon-Garber found that the older, founding members were loath to surrender power to the young. However, on her *kibbutz*, even these hardy pioneers, self-selected for a life of labor and danger, were shifting their interests from occupation to family. Harold Wershow (1969), who studied the same *kibbutz* eight years later, found less distress among the founding fathers: They had completed the transfer of power, they were resigned to it, and in their eighth decade they were beginning to enjoy a more family-centered life. In sum, older Israelis tend to withdraw from arenas of bold action, to concede this ground to younger men, and finally, after inner and outer conflict, to draw their pleasures from the present moment, in the more domesticated *kibbutz* setting.

A study by Joseph Cronin (1982) of American divorce lawyers shows a similar modal shift by age in the themes of professional practice, in a special sector of the legal profession. Cronin found that most of his younger subjects defined their role as divorce lawyers in adversarial terms. Approximately half of them were interested mainly in defeating the opposing lawyer, while the other half saw the legal system itself as their opponent. Manipulating the complexities of divorce law, they tried to extract the best possible settlement for their clients. But the senior lawyers see themselves primarily as marriage counselors. They are more nurturant, more interested in saving the marriage, "for the children's sake," than in wresting a victory from the opposing lawyer or an advantageous settlement through the legal code. When we consider that the senior lawyers had been socialized into the profession at a time when marriage counseling did not exist in a formalized sense, we realize that the shift toward a more giving, preserving form of legal practice reflects personal choice, seemingly based on later-life development, rather than the effect of job training and social pressure. There are no agreed-upon rules that mandate adversarial stances on the part of young divorce lawyers and "therapeutic" stances on the part of old divorce lawyers. The age-graded assortment of lawyers into these special roles speaks again to the force of developmental nature over social nurture.[8]

From Warrior to "Woman"

In brief, men who were once adversarial, whether as warriors, slash-and-burn agriculturalists, passionate pioneers, *politicos*, or trial lawyers, routinely become more pacific in later life; they turn to preserving life rather than killing, to maintaining social stability rather than fomenting ardent rebellion. In some cases, these transformations involve more than a drift away from flamboyant aggression, but an actual shift in gender distinctiveness, from univocal masculinity to sexual bimodality, or even implicit femininity.

Paradoxically, these gender reversals are most evident in those culture areas that also sponsor the fierce machismo of younger men, for example, rural Mexico and the American Southwest. Thus, as Indian women age, they can enter the ritual dances that are closed to younger women. By the same token, without feeling shame or censure, older Indian men can join the ranks of the women. Carl Jung (1933) tells of an Arapaho warrior-chief "to whom in middle age the great spirit appeared in a dream. The spirit announced to him that from then on he must sit among the women and children, wear women's clothes, and eat the food of women. He obeyed the dream without suffering a loss of prestige." Clearly, the old chief had legitimized his own wish, to become like a woman, by uttering that wish in the voice of the great spirit. Furthermore, his tribesmen did not find it strange that an old chief, thus legitimized, would join the women, just as they did not find it strange that the old women would join the men in ritual dance.[9]

Just as the old chief had rationalized his wish to be female, putting his unacknowledged desire in the mouth of a god, so some older men may hold their wives responsible for their hidden wish to engage in women's activities. Thus Powers (1877) writes of the Pomo:

> When an Indian becomes too infirm to serve any longer as a warrior or hunter, he is henceforth condemned to the life of a menial and scullion. He is compelled to assist the squaws in all their labors—in the picking of acorns and berries, in threshing out seed and wild oats, making bread, drying salmon, etc. These superannuated warriors are under the women's control as much as children and are obliged to obey their commands implicitly.

The general pattern is that older men are increasingly like women and with women. Becoming more domestic in their interests, they also retire into domestic space, the inner space of women. Thus Buell Quain (1948), reporting on the ferociously warlike Fiji Islanders, found that even though older men were more venerated than older women, they still "returned more closely into the seclusion of their households." While ceremonial activities may actually claim more of their time, their preference is clearly to spend their leisure among the women of their own household, who serve them coconut milk and honor their commands. They also spend long hours affectionately tending their gardens and remembering their ancestors. These erstwhile warriors become more "domesticated" and less interested in the company and affairs of men, and their lingering concern with productivity now retracts to the confines of their household gardens.

95

This feminization of older men is alluded to more explicitly by Leo Simmons (1945), who notes that aging Arawak warriors sometimes chose to cook for their families, despite the fact that such labor earned them the title "old woman."

Male Pacification and Male Pathology: Some Clinical Evidence

Until recently, few psychoanalysts studied the aging psyche; those who did portray the inner life of the older patient in ways that are strikingly congruent with the picture developed through our transcultural review of normal elders. Though he understood these phenomena in concretistic and quantitative (rather than developmental) terms, Jung (1933) stated with great clarity the point of this chapter:

> We might compare masculinity and femininity and their psychic components to a definite store of substances of which in the first half of life unequal use is made. A man consumes his large supply of masculine substance and has left over only the smaller amounts of feminine substance, which must now be put to use. Conversely, the woman allows her hitherto unused supply of masculinity to become active. This change is even more noticeable in the psychic realm than in the physical. . . . very often these changes are accompanied by all sorts of catastrophes in marriage, for it is not hard to discover what will happen when the husband discovers tender feelings and the wife her sharpness of mind.

As this statement suggests, Jung dabbled in alchemy. Hence the allusions to "masculinity" and "femininity" as substances that might be burnt off in a laboratory retort. But though he misconstrues the underlying developmental process, Jung's clinical observations are very accurate. Thus, in our own treatment with patients seeking psychotherapy for the first time in the later years (at the Northwestern University Medical School), we see many men who are shocked to discover their own "tender feelings," and many women who are shocked to discover their own "sharpness of mind."

In other settings as well, clinical work with American patients leads to psychodynamic formulations consistent with those independently derived from the study of normal aging in various parts of the world. Thus psychiatric observers—Joost Meerloo (1955), Norman

Zinberg and Stanley Kaufman (1963), and Martin Berezin (1963)—stress the aging male's withdrawal from active engagement with the world, in favor of more cerebral, introversive, even egocentric positions.

The general psychiatric position is best stated by Wolff (1959), a psychoanalyst who believes that late-onset disorders are activated by the reduction of male competitive drives in the later reaches of life. Coupled with this depletion, the return of infantile, id drives leaves the older man particularly prone to feelings of shame. Because of the same slackening of competitive drives, he cannot compensate for these regressive wishes through real or token victories, and is left feeling humiliated. These converging influences foster inferiority feelings, which in turn give rise to the endemic depressive episodes of later life.

Wolff suggests that the decrease in male aggression leaves an emptiness at the core of self, and that the resulting void is filled by depression. But I am in agreement with Jung, as cited above: When men use up their masculine aggression, they are freed to discover a potential resource, their hitherto unclaimed "feminine" substance. In the subsequent chapter, I present evidence for this "greening" of older men: Instead of pathology, reclaimed potentials, toward new growth, can emerge to fill the psychic gap left in men by the phasing out of powerful aggressive motives.

5

THE SEASON
OF THE SENSES

Old now, I know I'm no more use;
I've come home to a life of ease.
So many years I've slept in strange rooms,
Today I love my little house.
When I was out in the world, I hatched no great
   schemes;
At leisure now, I read old books.
As long as I can eat my fill,
I won't ask for anything else.

—TAI FU KU, 1167–?*

In the last two chapters, we tracked, through projective tests and ethnographic accounts, men's inner and outer migrations, away from competitive motives on the internal scene, and away from warlike, predatory, or entrepreneurial roles in the external, social world. But nature, including human nature, abhors a vacuum, and we see newly emerged, hitherto-hidden talents and potentials—toward the milder pleasures of hearth and home—showing themselves as the hot tides recede. As thanatos, the urge toward destruction, toward self-aggrandizement at the expense of others, leaches out of the male psyche, it reveals a previously undeveloped eros: a hidden capacity for cherishing, appreciating, and bringing together. In the preceding chapters, the focus was mainly on thanatos and its fate in the later years; in this chapter I will turn to the manifestations of eros, the new sensual powers that move forward to claim the territories abandoned by retreating thanatos.

* From Kojiro Yoshikawa, *An Introduction to Sung Poetry*, trans. B. Watson (Cambridge, Mass.: Harvard University Press, 1967). Copyright 1967 by the Harvard-Yenching Institute. Reprinted by permission of the publisher.

Aging and the Routes to Pleasure

The format and intent of our interviews has been discussed in chapter 2. Those procedures resulted in a total of 340 completed interviews from Mayan, Navajo, and Druze informants (including, in most cases, full TAT protocols). The interviews varied in length, depending on the degree of rapport, and on the subject's intelligence and mental vigor. In most cases, my field assistants and I got at least adequate data from each informant concerning life history; early memories; sources of pleasure, pain, and remedy; fantasy life (dreams and reveries); as well as health and social status. In this chapter I will continue to cite the informant's fantasies, as mobilized by the TAT, but I will also pay special attention to the self-report, interview data that bears on the eros of men—their conscious definitions of pleasure, beauty, and sustaining partnerships.

To recapitulate briefly, in earlier chapters we saw that the inner strivings of younger men, as registered in their private fantasies and public actions, are singularly aggressive in nature, aimed toward successful assertion of self, successful competition, and the productive, profitable manipulation of social and physical nature. Older men appear to abandon such forceful strivings and move to more inert postures, accommodating themselves to the aggressive initiatives of the great powers and authorities (including their aging wives) that people their worlds.

The open, exploratory interviews with Navajo, Druze, and Highland Maya men suggest that the loss of thanatos is part of a dialectic, in that the depletion of aggression is the necessary prelude to an advance toward new satisfactions. They also suggest that this shift along the mastery continuum is not limited to the inner life, to unconscious drives and covert imagery. Fantasy does provide a nonbehavioral route for expressing potentially dangerous emotions, but it also provides a screen for rehearsing potentially risky actions before they are given real play, through limb and muscle, in the real world. Fantasy is a prelude to as well as a substitute for behavior; it eventually fleshes out conscious thought, self-awareness, and purposeful action. Working on the assumption that inner needs, passive and active, registered in projective imagery would also shape more conscious, overtly expressed preferences, I routinely asked my informants to describe their major pleasures, and I devised categories that covered the spectrum of possibilities generated by their responses. Thus, in table A–7, shown in the appendix, the category Active Productivity refers to all forms of contentment, from Druze, Navajo, and Highland Maya sources, having to do with pleasure in work and in the capital produced by one's own labor. A Druze might say, "I am happy when the apple crop is good"; a Navajo might say, "I am happy when we have good grazing, plenty of good sheep and horses"; and a Highland Maya might say, "I am happy when I have my work, my wage, when there is plenty of corn and beans for me and my family."

When a subject tells us that his major source of contentment comes from domestic life, the presence of healthy, happy children, and an affectionate wife, his response is put under the heading Domestic Sentience: "I am happy when there is quiet at home, my children are healthy, and there is peace in the village. Then my mind is at rest" (Galilean Druze). A Navajo reports that he is happy when his relatives visit him; a Highland Maya says, in similar vein, that he is content when there is peace in the village and harmony in his house.

The Active Pleasure category accommodates responses from those informants, regardless of culture, who speak of travel, vigorous movement, sightseeing, and exploration as pleasures in their own right. A Druze might describe his pleasure in riding horseback, while a Navajo might speak of the pleasure that he feels when his relatives take him in their pickup to the Indian powwow at Flagstaff. In all cases, these pleasures are not passively received but are generated through action and actively sought out.

When the subject tells us that the sources of his contentment are ultimately out of his hands and that they involve a quiescent, waiting attitude on his part, plus an active stance on the part of some provider, then he is coded for Passive Receptivity. The pleasures of these men have to do with the consumption, the taking in, of pleasant sights, sounds, or substances that are produced for them or brought to them by others. They do not have to move actively to procure these gratuities. Thus a Druze says, "I am happiest when I am brought my first cup of coffee in the morning," or, "I am happy to have visits from my children, and good, respectable guests come to visit me." A Navajo: "I am happy when my relatives come back from the Squaw Dance, and bring me the big roast rib"; Highland Maya: "I am happy when there are soldiers in the village, to keep peace," or, "I am happy when they give me pox [liquor] and there is marimba music; we get happy then, and sing."

Figure 5.1 shows the distribution of respondents by age and source

FIGURE 5.1
Sources of Contentment:
By Age (Navajo, Druze, and Mayan Subjects)

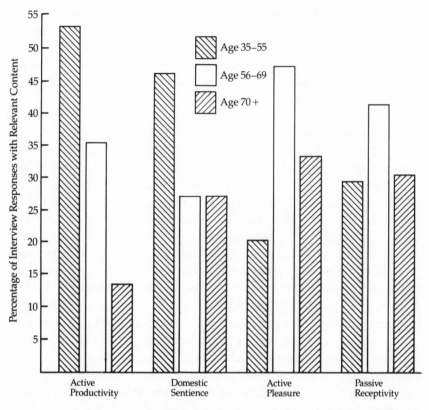

of gratification. The statistical significance of this distribution confirms the predictions based on prior TAT findings: Younger men are much more likely than older men to fix on Active Productivity as the primary basis of pleasure. But they also—and unpredictably—favor the domestic joys as well. In hindsight, we can see that the younger men concentrate on a mature, "parental" mix of pleasures: They feel at ease when their focused activity has a good productive outcome for themselves and their dependents, and they are also pleased when they have tangible evidence of their own productive value, in the form of healthy, happy children and contented households. In effect, the contentment of younger men rests on the two bases, love and work, that Freud declared to be pivotal to our happiness. Further, we infer that they can appreciate the "softer" domestic life because, for them, it is framed in a redeeming context of productive work—the

101

same work that takes them away from the potentially confining domestic life.

This balance of pleasures, between love and work, tends to hold through the seventh decade, though Passive Receptivity now acquires more importance for older men than the alternative, more active styles. For men in their seventies, the balance has shifted decisively away from work, away from agency and doing as sources of pleasure, and toward more sensual, receptive, and community-centered definitions of enjoyment.

When we inquire about the sources of pain and discontent, rather than pleasure, these findings hold up. More than half of the younger men (aged 35 to 54) blamed their discomforts on interference with their productive life: lack of rain (Navajo); loss of wage work (Highland Maya); expropriation of land, loss of their sons' labor because of Israeli army service (Druze). Younger men, then, complain of those frustrating events and agents that prevent the full use of their productive power. But with increasing age, the burden shifts: Increasingly, discontent is predicated on losses of those friends and kinsmen who provide material and emotional sustenance. The younger men complain when some outer agent prevents the full use of their own productive powers, but the older men complain when they are cut off from the presence, the goodwill, and the benign power of others.

Summing up these results, the younger men take their pleasure not only from aggressive action but also from being a source of security and provision to others, from the contentment and love of their dependents. But over the successive age groups, this position is undone and even reversed. As men get older, they draw less pleasure from being a source of security and provision to others through their own productive efforts; their pleasure comes to be based more and more directly on the productive efforts of others in their behalf, on the affections of others toward them, and on the satisfaction of their own appetites for tasty foods, pleasant sights, and soothing sounds. The appetite for production drops off in later life at a rate that parallels the cooling down of aggression. Power always has two faces, constructive and destructive; as the older man loses the sense of inner power, he no longer sees himself as a possible center of destructive forces, but by the same token he no longer presents himself as a center, a fountainhead of constructive energies. More and more, he draws his psychological sustenance from his own receptors—mouth, eyes, skin, ears—as well as from outer sources of strength: the reliable providers of satisfaction and security who cater to his senses and to his wish for security.

While these distributions are derived from the interviews, from a data source reviewed independently of the TAT, nevertheless the age shifts in the definition of pleasure, from active to more receptive forms of gratification, are quite consistent with predictions based on age shifts in the TAT data, from Active to Passive and/or Magical Mastery. To repeat: The male aging process is not simply a tale of losses; instead, as eros comes to inform behavior, attitude, and fantasy, older men discover (or rediscover) the varied pleasures of the senses, and of the warm, human relationships that they had earlier tended to subordinate or even avoid. If aging men lose some capacity for work, this loss may be partly compensated, as part of a dialectic of the appetites. Milder passions, those having more to do with community than with agency, collaboration rather than competition, receptivity rather than production, take over and energize the limbs and sense organs that have been abandoned by the more assertive drives. Aging men may lose some of their pleasure in winning, in imposing their will and their design on nature and other men, but this loss is partly compensated and turns out to be the precondition for a potential gain. All losses can be liberating; because they lose some pleasure in active striving, older men are by the same token freed up to discover new pleasures, of the sort that have little to do with either combat or production (and are even antagonistic to them).

The Season of Oral Pleasure

The relocation of pleasure sources in later life, the rediscovery by older men of early ("pregenital") gratification, is best illustrated by my findings concerning orality. The increase in receptive attitudes, as registered in TAT responses and interviews, alerted me to an increase in oral-receptive appetites, since the two, attitudes and appetite, are theoretically linked. Consequently, I judged that the age-shift in male psychology toward passive-incorporative modes would be matched and underwritten by a corresponding shift to the oral zone as a preferred site of pleasure.

I was gratified to find in the interviews that, across cultures, older men, much more than younger men, cast their depictions of joy, sorrow, and consolation in "feeding" or "oral" terms. Regardless of culture, happiness for older men equaled a good meal, and trouble seemed to be equated with lack of food or unreliable sources of supply. As my repertoire of cultures increased, I was impressed by the degree to which elders' accounts of the past as well as the present were figuratively stuffed with food: Pleasant memories recorded feasts or the meals that mother made. Unpleasant memories featured famine, the stinginess of stepmothers, or the greediness, at table, of siblings. My older informants were found in all manner of country—high desert plateau, rocky mountainside, fertile valley, and lowland scrub—and their social condition ranged from poverty (by local standards) to affluence and community leadership. These older men shared no common condition of social or physical place, no externalities that could account for their universal interest in food. And while they did have in common their relatively advanced age, the cultural meanings that attached to old age also varied across communities; their oral appetites could not be an expression of parochial age-graded requirements. Clearly, the food-centeredness of old men was more likely connected to "internal" demands, rather than to the external or social pressures, of later life.

Because of their unexpected association with age and their developmental implications, my older respondents' "oral" statements took on the status of data. In order to quantify and test the impression that such statements increased in number and intensity with age, I developed a coding system that accommodated all references to food production, preparation, and consumption, whether referring to the past or the present, whether found in the interview proper or in the projective protocols. The total orality score (TOS) is the sum of all these oral references.[1]

Having coded all Druze and Navajo data for these dimensions of orality, we discerned some interesting relationships between two orality subscores: the externalized projection score (EPS) and the syntonic orality score (SOS). Figure 5.2 indicates that EPS, which captures the tendency to experience oral longings in externalized, alienated form (as in the TAT figures), is high for younger Navajo and Druze, those in the age range 35–49. Indeed, for both these cultural groups, EPS is the highest score in the "oral profile" of the younger men. However, EPS drops off rapidly, and at about the same rate in both cultures, with increased age, indicating that externalized or alienated

FIGURE 5.2

Comparison of Navajo and Druze Externalized Projection Scores

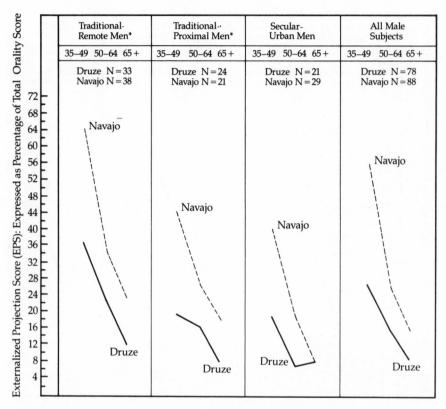

* Traditional-remote men are followers of the Navajo or Druze religions who live in unde-veloped, sparsely populated regions. Traditional-proximal men are old believers who live in native communities close to main roads.

expressions of orality occur proportionately less frequently, in the protocols of older Druze and Navajo men, than other, more directly personal avowals of oral need. In effect, the younger men of both these unrelated groups are telling us, "*Other* people—though not I— are hungry and needful" (and, by extension, "It is my job to feed them").

However, the older men, Druze and Navajo alike, in effect replace EPS with the syntonic (comfortable or acceptable) orality score (SOS),—which measures the tendency to value eating and to ac-knowledge appetite. As shown in figure 5.3, the SOS goes up, for both groups, at a rate directly commensurate with the decline of EPS. Most

FIGURE 5.3

Comparisons of Navajo and Druze Syntonic Orality Scores

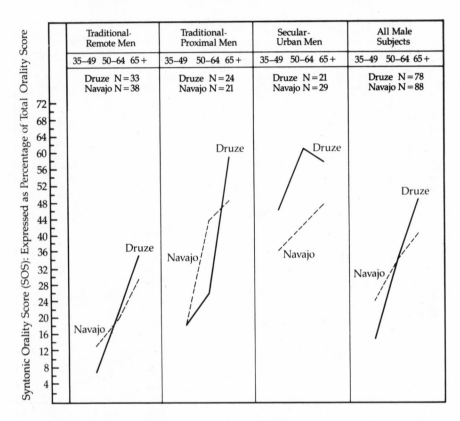

important, the mean SOS for both the Druze and the Navajo groups increases with age at a strikingly similar rate across these two very dissimilar societies, composed of people who could not have copied each other's ways. The correlation between EPS and SOS is not an artifact of our measurement strategies; EPS and SOS are independent measures, and the influence of age on both these variables is highly significant, as measured statistically. In both cultures, age alone accounts for the drop in EPS, and for the rise in SOS.

Clearly then, social pressures do not account for the dramatic age increase in oral definitions of pleasure among Druze and Navajo traditional agriculturalists. In both these drastically different societies, the younger men appear to disavow their oral needs, experiencing their own hunger externally, as a property of those who depend on them for their sustenance. The older men reverse this polarity: Playing down the awareness of oral need in others, they instead announce their own appetites, their own *pleasure* in the table. In both Druze and Navajo cultures, it appears that the older men no longer bother to dissemble or distance their own orality by experiencing it in others; rather, they *own* their orality, as part of themselves, and admit quite openly that it is a major pivot of their need and pleasure. If the younger man's implicit statement is, "Others are hungry, and it is my job to feed them," the older man's response in this implicit, cross-generational dialogue reads, "I need food for pleasure and security, and I largely depend on others to supply it for me."

In an earlier chapter, I showed that depictions of struggle, particularly between men, tend to disappear from the later work of creative artists. The same body of work demonstrates that more pacific oral content, depictions of food and eating, comes to fill the space vacated by the absent aggressive content. When we turn again to the work of George Price, we find that, as a young artist, most of his cartoons contained only a few references to food. But as he aged, incidental references to foodstuffs, food preparation, and eating began to appear, and food is very often the theme or punch line of the most recent cartoons. Oral content increases in Price's drawings in step with the phasing out of the Man's World and the phasing in of the interior Woman's World. (Interestingly, when we plot the increase by age, of oral references in Price's drawings [fig. 5.4], we find that the slope closely parallels the age curve for syntonic orality scores developed from the Druze and Navajo materials. The elder Price makes cartoons out of the same hungers that he shares with most old men of our species.)

These findings prompt a question: Does the intrinsic push toward later-life orality point to a relatively unimportant phenomenon of later life—namely, the appearance of gourmet tastes among older men—or does it have wider implications, for a general psychology of later life? Our evidence supports the latter construction: The age-related rise in orality betokens more than an older man's enlarged interest in groceries; instead, it points to the arousal of passive, incorporative personality tendencies or modalities, possibly affecting a wide range of specific zones and activities, that go far beyond the pleasurable consumption of food per se. Thus Alan Krohn and I (1971) found that the syntonic orality scores achieved by Navajo subjects predicted to the ego mastery orientations expressed in their dreams. Subjects of any age, with low orality scores, tended to have Active Mastery dreams, featuring vigorous action on the part of the dreamer,

FIGURE 5.4

Age and Oral Content: In Navajo Interviews,
Druze Interviews, and the Cartoons of George Price

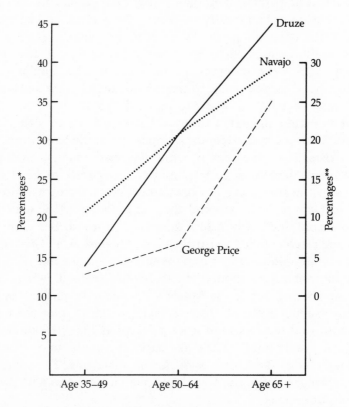

* Syntonic orality score (SOS) expressed as percentage of total orality score (TOS).
** Percentage of Price drawings with major oral theme.

usually in outdoor settings. Conversely, subjects of any age with high orality scores were much more likely to have Passive Mastery dreams, those in which the dreamer was either absent from or on the periphery of the drama, in which other characters were the main actors, and in which the dreamer, if actually present, was either inert or at the mercy of some arbitrary force: "I dreamt that some big animal was chasing me." Figure 5.5 charts this statistically significant relationship, one that points up the association between orality as zone, an erotic in-nervation of the mouth, and orality as mode, an attitude of depen-dency toward all sources of security and nutriment, including, but not exclusive to, foodstuffs as such.

108

FIGURE 5.5

Distribution of Navajo Dreams by Mastery Style and
by Syntonic Orality Scores

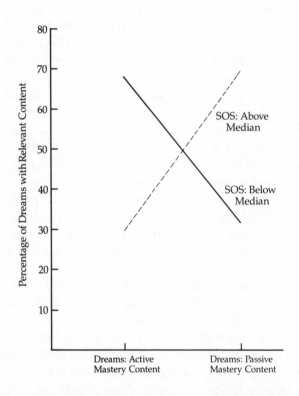

The association between orality and a more general passive orientation is not limited to dreams and fantasies, to visions of sugar-plums on one hand and dreams of being hurt or victimized on the other. Indeed, as measured on the SOS barometer, orality appears to be a most crucial dimension of waking experience in later life. For example, as seen in figure 5.3, the SOS predicts the community organization and location of Druze and Navajo subjects: Controlling for age, we find that remote-living, traditionalist Navajo and Druze will have orality scores that are significantly lower by statistical test than Druze and Navajo matched for age who live close to (but not in) urban environments (in the Navajo case, the tribal administration center of Tuba City; in the Druze case, the Jewish port city of Haifa). In other words, while variations in the male orality quotient are not affected by cultural content per se, they are responsive to social structure and social ecology—the forms, rather than the particular values, of communal association. Thus the socioreligious values of remote settings are not very different from those that obtain in more protected settings; rather, it is the internal receptivity to such norms and values that varies with social ecology. Cultural content per se is not mapped directly into personality; rather, personality is most affected by those social-ecological conditions that determine not the cultural ideas themselves but the degree to which these are assimilated by the individual psyche over the course of socialization.

Accordingly traditional men, dwelling in remote, intimate communities are more exposed to social censure and are perhaps more sensitive to religious injunctions against pleasuring the flesh through oral indulgence; the harsh conditions of their isolated life, plus the vulnerability of their dependents, put them in a condition of chronic emergency and make them sensitive to shaming and social censure (as by gossiping neighbors) against any appearance of passivity and selfishness. But even in these physically and socially rigorous settings, the level of male orality increases at the same rate with age as is the case in those settlements (urban-proximal) that are close to urban centers.

The SOS also predicts to the degree of Navajo alcoholism: Within age groups, the higher the orality score, the greater the likelihood that the Navajo subject will have a reputation for heavy drinking as well as a medical record for alcoholism. Note that the addiction to alcohol is not a direct extension of the "oral" interest in food; rather, alcoholism is mediated by depression, an emotional disorder that is a frequent side effect of the oral character. These alcohol-abusing Navajo do not seek strong drink because they like the taste or to quell hunger, but because their unsatisfied neediness makes them feel empty and deprived; the same oral character disposition leads them to seek remedy in any intoxicating liquors that function like mother's milk for an infant, bringing temporary relief from fretfulness.

Further evidence that the oral character contributes to symptoms that have no connection with food intake per se comes from the finding that syntonic orality scores calculated at Time$_1$ predict to the Time$_2$ health status of the Navajo respondents. Again (see fig. 5.6), regardless of age, Navajo with high orality scores at Time$_1$ are much more likely to show severe illness (as estimated from the Public Health Service records) at Time$_2$, than Navajo with low Time$_1$ orality scores. Men with high Time$_1$ orality scores who were not significantly ill at Time$_1$ are much more likely to have died or to have developed severe, even lethal, illness by Time$_2$.

FIGURE 5.6

Distribution of Navajo Syntonic Orality Scores
by Subject's Health Status

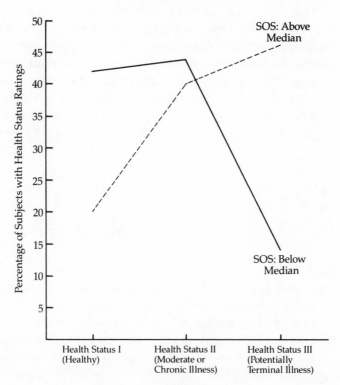

The significance of this relationship between orality and illness in the Navajo population is not clear. It may be that their oral tendencies make Navajo men more prone to depression and thus more vulnerable to the physical diseases that can occur when the body's immune systems, as a side effect of that same depression, break down. More likely, in the Navajo case, the heavy drinking that is allied to the oral character eventually weakens the body, thereby mediating between orality and severe illness. (In the traditional Druze population, where male boozing is against the religion and so almost non-existent, the relationship between orality and age is still found, but the relationship between orality and illness is much less evident.)

Finally, it may be that orality does not "cause" illness but is caused by it: The anxious, orally disposed subject, uneasily sensing the onset of severe illness, may turn to alcohol, an oral anodyne, as a tranquilizer. At Time$_1$, prior to the onset of clinically significant symptoms, certain subjects may have sensed, preconsciously, the oncoming lethal disease that flowered by Time$_2$ and may have sought out oral means of comfort. Faced with the ending of their life, they have perhaps revived "oral" forms of soothing, like the milk provided by their comforting mothers at the beginning of life. However, for the purposes of our argument, the flow of causality, whether from orality to illness or from illness to orality, is not the issue; more important is that oral investment, whether as cause or consequence, is not just a zonal property of the mouth and its mucous membranes but is also a modal property of personality, one that invests and determines a wide range of psychological activities, defensive as well as explicitly nutritional, in later life.

The Hunger of Old Men: A Human Universal

The orality score appears to be a valid measure of all those male personality modes that are centered on consumption, on comfort, and on the pursuit of security through accommodation rather than more active means. The SOS registers the emergence, in proper season, of the more self-indulgent, self-nurturing posture that replaces youthful aggression and productivity. Thus younger men place themselves in an active and protective stance, to meet the oral needs of others (and thereby to vicariously gratify their own); but older men are less likely to manage their orality by feeding others or by shunting their own needs into fantasy, into the stuff of dreams. Instead, they declare their hungers directly to the world and manage these needs by trying to influence potential providers in their favor. They either accommodate themselves to the requirements of authorities and gods, or, by translating their special hungers into sacred dogma, get their way through altering the ritual systems of society in their own favor.

Documenting the latter point, Simmons (1945) cites evidence from more than thirty widely differing preliterate groups. Across this diverse cultural range, older men have typically used their social prestige as a lever to ensure the choicest foods for themselves, by making them taboo for younger men. He notes that the "special assurance of food . . . becomes a matter of increasing concern in later life." And he adds, "As far back as we can go in human associations, the hands of the aged have reached out for food when they could find little else." His transsocietal review shows that "most primitive societies have conventions to regulate the allocation of food to the aged." In support, he provides anecdotal evidence from Asian and Amerindian groups: the Yukaghir of Siberia; the Chuckee, also of Siberia; and the Omaha, Crow, Kwakiutl, Chippewa, and Iroquois Indians of North America. In all these settings, youth who might be tempted to violate the food taboos that preserved the choicest foods for the aged were threatened with supernatural punishment.

My own literature search has turned up similar examples from other parts of the world, all of which suggest that their access to food is a clear index of the older man's social standing. The best illustration of this pattern in nature is provided by Waud Kracke (1977), who studies the Kagwahiv of the Amazon Basin:

> Kagwahiv who reach old age are highly respected, not only for their knowledge of tradition, but also for the spiritual strength that has permitted them so long a life. This is recognized in the relaxation of food tabus, for an old person is considered to have the spiritual strength to withstand the power even of such meats as paca and curassow.[2]

The high degree of association in the primitive world between advanced age, elevated social status, and gourmandizing is too widespread to be accidental. It tells us that the aged men of patriarchal societies have used their great power to secure their pleasures, and that, for them, as with their age peers in nonpatriarchal societies, pleasure is equated with food. In a gerontocracy the elders make the rules, and so the law of the land they rule reveals the innate preference of its elders. Again, it could be argued that, despite the wide-ranging association between gerontocracy and dietary laws that favor the aged, it is the law that creates the appetites, rather than the appetites that create the law. But among the Zanaki, as described by Otto Bischofsberger (1972), the aged love those foods that are also theirs by legal title. In this age-set society, the age grade of the old men disposes much power, and "millet is the staple crop to which especially the old people showed great emotional attachment." Clearly, elders do not grudgingly accept, as their social duty, the special nourishment that custom sets aside for them; among the Zanaki (and the Navajo and the Druze), they *lust* for the stuff.

The oral cravings of the elderly emerge even in settings that are not gerontocratic. The aged of Western societies share the same hungers as their more primitive peers, even though they lack the social power to legislate their gourmet preferences into law, custom, or taboo. Thus Jennie Keith-Ross (1974) discovered that Frenchmen who retired to an old-age home were "socialized" into their new abode via the dining hall:

> New friendships become public in the dining room, major battle lines are marked there, and location of the table represents location in the networks along which information is whispered or shouted during the meals. Quarantine at a table alone was a strong sanction of antisocial behavior which was applied, for example, to a man who frequently got drunk and insulted the women who lived next door to him.

In sum, all major appetites and concerns converged in the dining room, the arena of central need for that age group. And these oral priorities cannot be blamed on the special gourmet interests of the French. David Guttman (1973) finds parallel interests among the Jewish aged clustered in a community center in an eastern American city. There, the most popular program offered is the hot lunch: "In addition to the low charge, it seems to be meeting the social and nutritional needs of many aged."

But the sanitized language of the above citation avoids the central fact: Food does more than meet "social and nutritional needs." In later life, food is sexy, the erotic keynote of healthy later life. Excitement about food can ignite a full spectrum of political philosophies and political activities, even among the institutionalized old-old, when residents of an old people's home are given democratic jurisdiction over their menus. For example, a story in the *New York Times* reports on the lively consequences when creative administration works with the oral eros of old-old residents:

> Senators are elected every six years, Presidents of the United States every four and food items are voted onto and off the menu every two years at the Hebrew Home in the Bronx. Senatorial and Presidential elections are considered relatively inconsequential compared with what the residents of the home eat.
>
> "A bum Senator we can live with," said 1 of the 19 voting members making a speech at the meeting, where tasting, debating and balloting took place. "Ratatouille we cannot."
>
> "Nothing is more important to people here than meals," said Marcus Solot, a member of the Food Committee. "It's just about all anyone talks about." . . .
>
> There are many single-issue factions who want the ear of committee members, such as those seeking less sodium in the food, less sugar, and more spices. There is the "no breakfast before 8 A.M." group,

the al dente spaghetti coalition and the anti-cheese factions, including a "no-Parmesan" splinter group, and those pushing for larger portions. . . .

Mrs. Small said sometimes members think about leaving the Food Committee, but they cannot bring themselves to give up the power and prestige.*

These accounts tell us of a full, even vibrant social life, a cockpit of power struggles, of thanatos as well as eros, that is turned on by the "oral" intervention; they also tell us that such appetites are not exclusive to old men. Senior women are equally passionate, using their control over the oral resources as an avenue to political "power and prestige."

Incidentally, Jews, like the French retirees cited earlier, are reputedly a food-obsessed people, but these patterns appear to be independent of national character. Thus Roger and Louise Barker (1961) contrasted the "psychological ecologies" of aging in a Midwestern and an English town, finding that in both these relatively WASP settings, the older people preserve and even increase their interest in nutrition, even as their investments in other activities fade out.

Clearly then, the appetites of old men exist prior to any social etiquette: Their hunger has an imperative of its own. Since these findings by independent investigators, across a range of human societies, were predictable from my own findings of age-related increases in SOS among the Navajo and the Druze, we must concede that the special hungers of older men are culture-free. Despite varying cultural conventions regulating the distribution and value of food in later life, the rate of increase in oral concerns with age appears to be a cross-cultural constant, a universal. Not traceable to cultural nurture and leading to new satisfactions, this oral surge appears to be developmental in nature. Time does take away from older men their edge, their genital and phallic appetites for dominance, victory, and successful agency, but it gives back an extended range of hitherto closeted pleasures, those of the table and the community of *companions*, literally, those who take bread together.

* From William E. Geist, "About New York: Where Politics Is Thought for Food," *New York Times*, March 12, 1986. Copyright © 1986 by The New York Times Company. Reprinted by permission.

Orality and Death

While Simmons (1945) grants the primary and universality of oral preoccupations in later life, he does not see these as surgent, proactive, or developmental in nature. Rather, he sees orality as reactive, a remedy against death, against the great universal that marks the end of development:

> A dominant interest in old age is to live long, perhaps as long as possible. Therefore, food becomes a matter of increasing concern. Its provision in suitable form, on regular schedule, and in proper amounts depends more and more upon the efforts of those who are in a position to provide or withhold it. And, as life goes on, the problem of supplying and feeding the aged eventually reaches the stage at which they require the choicest morsels and the tenderest care.

In this calculus, older men do not love food; they fear death. Valuing long life, they naturally become diet-conscious and take food as a kind of medicine against death.

Simmons is wrong. For one thing, there is no reason to believe that the equation of food with health (a cultural matter) is as generally distributed as the orality of later life (which is universal). Cultures vary greatly in their awareness of the relationship between nutrition and longevity. Moreover, if senescent orality swells in fearful anticipation of death, then we would predict an overall elevation of the total orality score (TOS) by age, across cultures. But this is not the case: In point of fact, TOS stays relatively constant across Druze and Navajo age groups. What changes predictably with age is the manner in which oral concern is expressed. Age does not change the sheer amount of oral interest but rather the mangement of this interest and its expression. Younger men, though far from death, do not lack oral appetites; rather, they keep these out of direct consciousness by projecting them, and by satisfying them in and through their external representatives: dependent wives, children, and even livestock. Meanwhile, their own syntonic orality is waiting in the wings, against the time when they can turn over their responsibilities for feeding others and relax into the consumer's role.

Orality and Ego Defense in Later Life

Earlier we saw that oral pleasures serve as remedy against various discontents and insults of later life, against illness as well as dying. But it is not yet clear why oral pleasures, instead of others, should be selected to guarantee the sense of security in later life. Though unbuttressed by statistical analysis, some crucial case studies may provide answers. These suggest that very potent natural tranquilizers toward the end of life call up the climate of the beginning of life or early childhood. By literally reliving the weather of his beginning, the older person denies the approaching end. In any culture, the first pleasant experiences in life center on feeding, on the relief from fretfulness that feeding brings; this universal experience may determine the equally universal enlistment of oral pleasures in later-life attempts at magical reassurance or defensive denial.

The following statement made to me by a tough old Navajo medicine man (83 years old, then riddled with cancer and now deceased) illustrates the ways in which food can provide, at the end of life, the linkage to early pleasure and the consoling sense of communion with mother:

> Every time I eat the white man's bread—white bread—I eat plenty, but I don't know what happens to it. It doesn't stay down here. What I need is corn meal mush, dumplings. . . . I'm expecting my daughter—she comes around once in a while and sometimes she makes me that corn meal. . . . When I was a baby about a year old I found that they were putting that corn meal mush in my mouth. That's what I'm raised on, and I'm still hungry for it.

This mortally ill Navajo traditional is not lusting after corn meal mush because of its special nutritional benefits or its realistic role in fending off death; he remembers its taste, as well as the special climate of mothering in which that savor was first encountered. This maternal ambience is currently conjured up for him by "motherly" feeding from his daughter.

This use of special foods as the touchstone to early experience may be seen in old Caucasian women, as well as old Amerindian men. Sandra Howell and Martin Loeb (1969) tell of an elderly grandmother who refused food in her American children's home, claiming that the

117

only thing she really wanted was boiled potatoes and buttermilk, the primary foods that she had eaten as a child growing up in Eastern Europe over seventy years before. In later life, eating is a form of reminiscence; better yet, it is a way of actually reliving the past. And as men recycle the past through oral pleasure, they deny the present and its portents of their oncoming death. Again, the connection between food and reassurance pivots on the pleasurable feeling, not on some rational calculus that links nutrition to the prolongation of life.

The distinction between Simmons's view and my own is made clear by the following dream, reported by a 65-year-old informant of Aguacatenango (Highland Maya). Earlier, Don Cyrilo had told me that his death was near, and his subsequent dream associations dramatize the special role that food plays in dealing with such threats:

> I dream that they are cutting meat and that they give me the meat as a gift. . . . I run from something. I escape through a crack in the wall. A man tries to catch me, and I can't run. Then, suddenly, I fly away. A woman calls me but I don't want to go to her. I am offered food. They offer me grains of corn. I put it in a pocket and eat it.

Here we see, without much camouflage, the threat of death: the assassin, the man who implacably pursues. But the terror of death is countered by the promise of oral pleasure, by oral phrasings of denial: Pursued by death, Don Cyrilo magically flies away, and he is offered gifts of meat and grain. The same recruitment of orality to the service of denial (rather than life-extending medication) came out in Don Cyrilo's discussion of his heavy drinking: "When I drink for two days, I don't feel alive; it's just like flying." Thus strong drink makes Don Cyrilo forget his troubles, and it gives him the wings that, as in his dream, speed his flight from the assassin. Again, it is the irrational, magical meaning of food and drink that turn them into remedies against the fear of death, not their objective nutritional or medicinal virtues.

Home to the Mother

Don Cyrilo dreams of the assassin, but also of a beckoning woman. In the context of this death-haunted dream, we sense that this is a maternal presence, one of the comforters that Don Cyrilo calls up, along with food and magic wings, to counter the shock and pull of death. In this dream, the mother is linked with meat and grain as a comforting presence; the oral and the maternal are confounded, and for good reason: As archaic appetites are revived, we can expect that the ties to the parental figures involved in satisfying these early hungers will also return. When the oral demands of early life are rekindled, older men will, like the cancer-ridden Navajo medicine man, reinvest their early memories of the nutritive mother, they will seek out maternal women in their present social surroundings, and they will even treat younger women, such as their daughters, as though they were actual mothers.

Transcultural and other empirical evidence for a revived maternal investment among older men is patchy but suggestive. As shown in figure 5.7, there is a tendency (not statistically significant) for older Kansas City and Highland Maya men, when asked, "Who was the most important person in your life when you were a young child?" to respond, "My mother." Younger men of these same groups are more apt to cite their fathers as the formative influences in their childhood.

A similar trend is noted among the most traditional and remote-living Navajo group: As shown in figure 5.8, the percentage of spontaneously given maternal memories increases markedly with age, while spontaneous memories of fathers and "father figures" decreases at a corresponding rate.[3]

An equivalent "mother-seeking" tendency also shows up in stories told by the oldest Mayan men, both Lowland and Highland, in response to a TAT card showing a woman who holds a nursling and looks toward two children, while these tend a flock of chickens. As seen in figure 5.9, the younger men focused on the productive or domestic implications of the scene: The family as a whole were enjoying their flock, or the mother was instructing the children in the care and feeding of chickens. But eight of the oldest Maya (and none of the younger men) dealt exclusively with the mother and her baby, while neglecting to mention either the older children or the poultry. "The baby is crying—it wants to nurse" or, "The mother is concerned that her baby might be sick," were typical stories. Productivity be damned: The older Maya want their mothers.

Across cultures, the seeming shift to more oral-incorporative appetites seems to bring about equivalent shifts in the social preferences and relations of later life. Memories about early caretakers, those prominent during the oral period of development, are revived and

FIGURE 5.7
Age Distribution of Kansas City
and Highland Maya Responses to the Question:
"Who Was the Most Important Person When You Were a Child?"

FIGURE 5.7
Age Distribution of Kansas City
and Highland Maya Responses to the Question:
"Who Was the Most Important Person When You Were a Child?"

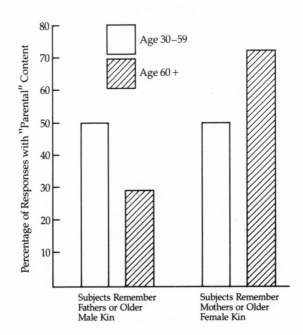

reinvested, and these idealized memories are mapped onto the current social scene, directing the choice of intimates (and enemies) within that field: Women, even much younger women, are viewed as though they were nutritive or ungiving mothers, and male contemporaries are sometimes related to as though they were envious siblings, the greedy, rivalrous children of these same mothers.[4]

Self-Consolation as an Ego-Defense:
The Reliance on Denial

While older men may yearn for their mothers, they may partially satisfy that unrequited longing by becoming the lost mother, not only toward others but also toward themselves, and in ways that can alter

FIGURE 5.8

Traditional Navajo Age-group Comparisons: Percentages of Memory
Materials Relating to Mothers and Fathers*

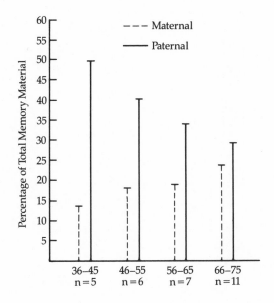

* For this chart and the work that it represents I gratefully acknowledge the efforts of Dr.
Jeffrey Urist.

their experience of the world around them. As the maternal bond is
revived, the older person can take over, for himself, the mother's
functions of making trouble go away; as he magically dispels the
awareness of threat and anxiety in his current life, he enacts the ego
defense known as denial.

In the more formal sense, denial represents the tendency to confound wish with reality and to replace troubling realities with their
opposite: The bad is magically transformed, by perceptual fiat, into
the good. Given the "oral" priorities of later life, we expect that the
tendency to denial would increase in older subjects and could be observed in the Magical Mastery responses to those TAT cards that
portrayed threat or trouble. We should find that, with advancing age,
realistic stories about such distressful cards are decisively ignored in
favor of interpretations that deny the troubling card implications and
replace them with more soothing, self-comforting ideas and images.

121

FIGURE 5.9
The Family Scene Card:
Distribution of Lowland Maya Responses by Age and Theme

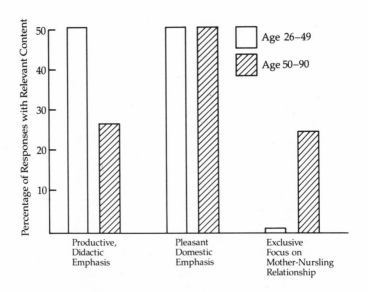

These assumptions have been tested via two special cards, used in their unmodified form at all sites (except Kansas City). Both the Desert Scene and the Bats and Man cards suggest grim possibilities, of the sort that are likely to stimulate self-soothing responses in vulnerable individuals. The Desert Scene (fig. 5.10), a TAT card first prepared for an American Indian population, depicts a rather barren desert landscape, empty of people, spotted with a few shrubs, cacti, and the skull of a cow. Man and his works are represented by a fence that bisects the scene, as well as by the hint of a trail. In one sense, the scene represents bleakness, desolation, and uncaring nature. In another sense, it represents the dynamic opposition between the entropic forces of the universe, ignorant of human needs, and the ordering hand of man—as represented by fences, roads, and cattle—which attempts to shape raw nature to human purposes. The various mastery positions devised for this card are reported in the appendix, under "The Denial-eliciting Cards." They were intended to capture the ways in which this opposition, as well as the theme of essential barrenness, is managed by the individual respondents.

Briefly, younger men are more likely to acknowledge, without

FIGURE 5.10
The Desert Scene Card

denial, the pictured reality of an ungiving world. However, as men age, images of bounty more and more come to replace the harsh realities of the original scene. Instead of unrelieved dryness, rain falls from the background clouds or sluices through the foreground arroyos. Instead of cacti, sprouting crops green the dunes. Finally, for the oldest men, sheer invention takes over: Fence posts become angels, the foreground gulley is given the shape of a woman, or the background clouds become bags of money. For the oldest men, the forms that deny reality become the perceived reality.

Similarly for the Bats and Man card (fig. 5.11), based on a Goya etching, *The Dreams of Reason*. It shows a recumbent man under a swarm of demonic, batlike creatures, and suggests the dangers of undefended passivity. Younger men are more likely to acknowledge the threat inherent in the scene, and even to suggest that the man could strike back against his tormentors. But once more, the older men tend to change threatening stimuli into their benign opposites. The winged demons become mere birds, the man's head becomes a bush, and the

123

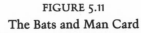

FIGURE 5.11
The Bats and Man Card

whole scene becomes one in which birds drop toward their nest, rather than stooping on their prey. In some versions given by the oldest men, the reversal is complete: They see the demonic birds as angels (see the appendix, under "The Denial-eliciting Cards").

These Pollyanaish interpretations from older men do not represent the random errors of dementia; these are motivated misperceptions, designed to replace troubling stimuli with soothing, consoling experiences. Perception has, in effect, come under the sway of the eros of later life, the eros of the eyes. In later life, the eyes not only collect the available visual bounty; in addition, the visual apparatus actively reworks the environment, converting, by a kind of perceptual fiat, harsh realities into pleasant experiences. In effect, the unmediated wish for pleasure, rather than the flinty realities of the outer world, becomes the arbiter of perception. The eye sees the wish (more precisely, the wish-as-realized) instead of the troubling features of the world.

Restoring the Mother: The Aging Artist

Earlier we saw that older artists, like George Price, turn the oral hungers that are common to older men into the stuff of their art. Similarly, the maternal yearnings that are intimately tied to the oral complex are also captured in paint by great artists, in their proper season. We have already seen, in the chapter on male aggression, that in later life older artists replace images of active men in outdoor space with images of powerful, sumptuous women filling interior, domestic space. Thus, in typical works of William Steig's later years, women are portrayed as little more than one grand sweep of bosom, with a diminished though still regal head perched atop this wealth of milky flesh. In effect, the artist has recaptured a child's-eye view of the mother.[5]

In such drawings these female paragons are usually attended by rather insubstantial *Luftmenschen*, who court them from a distance or, as bums and beggars, leer hopelessly at them through the bars of iron gates. For example, in a Steig drawing appropriately titled *Queen Bee*, an ample woman in a flowered dress is attended by three adult but shrimpy men, no taller than her hips, who cluster around her skirts like children around the mother.

Some older artists still keep a modicum of distance from the woman/mother. Even within the intimacy of a sealed room, Edward Hopper's men and women look past each other: Social distance compensates for the loss of physical distance. In Pablo Picasso's case, we see in his final paintings the abolition of all intersexual distance. Thus he concentrates on a sequence in which the painter and his female model are interfused, in such a way that each creates and contains the other. And in his final cycle of *Baiser* (*Kiss*) paintings, Picasso presses the male and female together, joining them by a common mouth. In this frantic clinging, limbs and parts lose their gender: Breasts become the base of the phallus, and the phallus becomes part of the woman that it penetrates. In these conceptions, we see the final entry into female space, and perhaps the ultimate fantasy: union with the mother.

Mouth Pleasure and Eye Pleasure

Our contention that the oral hungers of old men are developmental in nature and have a species-instinctual rather than a rational basis is further supported by some unexpected findings concerning other, belated expressions of sensuality, those not related in any causal or logical sense to nutrition or longevity. Especially striking is the way aging men in different cultures increasingly come to use their eyes as registers of the world's beauty rather than as instrumental collectors of useful information.

This trend toward ocular sensuality is graphically revealed when we classify the references to visual activity in responses to the Rope Climber card. All references to "seeing" or "being seen" on the part of the climber were arrayed, blind for age, on a continuum, from the most active and instrumental use of the eyes to their most aesthetic, sensual, and receptive employment. A distribution of such entries by category and age shows that younger men tend more often than older men to propose that the hero is looked at rather than looking. Younger Americans in particular see the climber as the center of an audience's admiring attention: "He's showing off his strength—the audience gets a thrill out of looking at him." According to these younger Americans, the rope climber produces visual bonuses for a receptive, appreciative audience. By the same token, younger men, when they mention visual activity, imply that the eyes are in the service of the instrumental act, guiding the climber toward his goal. But for older men, looking itself becomes the goal of action: "He's climbing to get a good view," or, "He's resting on the rope and looking at something beautiful." Some instrumental use of the eyes is exchanged for a pleasurable, even erotic, function (so much so that some older Kansas City men—but no younger men—saw the Rope Climber as a sexual voyeur).

The same age trend toward heightened pleasure in looking and seeing emerges in responses to the Boy on Cliff card of the TAT battery (fig. 5.12). This shows a younger male figure (dressed in breechcloth for Indian respondents and in long pants for Middle Eastern respondents) standing atop a cliff or mesa and looking out over a possibly barren, desert landscape. While the age trends toward aggravated visualism are not picked up by this card in all cultures, in the case of the Navajo the predicted age differences are quite clear (see fig. 5.13).

FIGURE 5.12
The Boy on Cliff Card

The younger Navajo typically propose that the hero is using his eyes in the service of production (looking for new crop lands, checking on his present cultivation), mobility (searching for a good route through rough country), or warfare (scouting for enemy forces). For the older Navajo (and, to a lesser extent, the older Maya), the looking is eroticized, pleasurable in itself. Although the landscape presented in the TAT card is virtually featureless and normally suggests dry, barren country, a significant number of older Indians hold that the hero has come to the high point to look at beautiful scenery, frequently an ocean. (In the appendix, table A–11 provides a breakdown by culture, age, and response category.)

Figure 5.14 sums up the aggregate ocular-erotic trend for all cards. Across the expanded sample, the noted pattern holds. Younger respondents emphasize in their stories the functional, agentic use of the eyes, as auxiliaries to productive action. Older men, across societies, propose that their eyes garner pleasure or that looking without practical purpose is an end in itself, indulged for its own sake. This

FIGURE 5.13

The Boy on Cliff Card: Distribution of Navajo and
Mayan Responses by Age and Theme

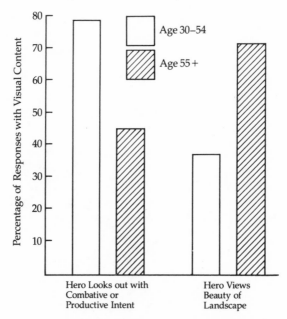

shift, from purposeful to purposeless ocularity, is statistically significant at beyond the .005 level.

The same age trend, toward eye pleasure, organizes some portion
of the Rorschach responses given by the Navajo. Those responses
were noted that made specific reference to beautiful objects, colors,
and landscapes, and the percentage of such responses was computed
for each individual record. The mean percentage of such responses
was then computed for the youngest (35–49) and the oldest (60 and
over) Navajo groups. The difference between these means was significant at the .05 level.

Visual eroticism shows up in behavior as well as fantasy. Robert
Kubey (1980) has reviewed the work of numerous investigators, who
generally report that the elderly are prime consumers of television,
particularly public affairs broadcasting.[6] And while it could be argued
that the retired aged turn to television out of inertia rather than active
desire, it is significant that elderly people in the United States were
predominantly "visual" in their interests long before television ex-
ploded into our daily life. Thus, in the 1930s, the Strong Test (1931) was
used to measure age differences in leisure interests in an American

FIGURE 5.14

All Cards: Age Distribution of TAT Responses with Visual Content
(Navajo, Druze, and Mayan Subjects)

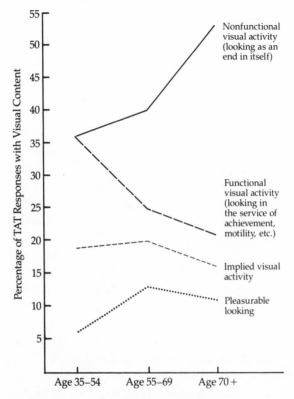

middle-class population. The two activities that most increased with age were bird-watching and visiting museums: Old people take their eyes for an outing.

Again, while this late onset of visual sensitivity cannot be understood in rational terms, as a reasonable reaction to later-life changes, it is predictable from psychoanalytic theory, which lays stress on the irrational side of man. Thus, for Otto Fenichel (1942), oral and visual pleasure strivings are organically if not logically linked, in such a way that oral eroticism predicts to scoptophilia (pleasure in looking), and it is not accidental that both kinds of satisfaction move to psychic prominence in later life: "The eye may represent pregenital erogenous zones symbolically. As a sense organ it may express oral-incorporative and oral-sadistic longings in particular. . . . Such 'oral' use of the eyes represents the regression of visual perceptions to the incorporation aims that were once connected with early perception in general."

In short, from the psychoanalytic perspective, the activity of the ocular zone comes under the dominance of the oral-incorporative mode: The older man "eats" his world with his eyes.

The Omni-Pleasure Principle of Later Life

The trend toward visual sensuality, then, may be taken as only one indicator of an organic and wide-ranging psychosexual development of later life: the general, tidal sweep of the erotic life back toward its dispersed, diffuse origins. This is not primarily a return to orality per se; instead, the increases in oral and visual pleasures are two equally dependent events brought about by an independent reassertion of the omnimodal pleasure principle in later life. Paradoxically, this insight into later life rests on psychoanalytic developmental theories, proposed by Karl Abraham (1953), to cover the course of early libidinal development. In infants, he held, all bodily functions are exercised for the erotic pleasures they afford. But, later maturation requires that the erotic goals of the individual receptors be suppressed so that they will no longer be distracted by immediate excitements, and so that their functions, rendered steady and reliable by these developments, can be put to the utilitarian service of the ego, its instruments, and its practical goals. The erotic strivings do not finally disappear but are, in the course of normal development, boundaried and contained in the genitals. In effect, the diffuse but persistent pleasures of early life are traded for the infrequent, spasmodic ecstasy of the orgasm. Thus the early, diffuse erotic strivings that originally serve individual pleasure are ultimately subsumed to larger procreative and species goals.

Concerning this splitting away of pleasure functions from the receptor zones, Sandor Ferenczi (1913) has written: "If there were no such separation of pleasure activities, the eye would be absorbed in erotic looking, the mouth would be exclusively employed in necessary self-preservation activities."

In short, the growing individual, who can no longer depend completely on the instrumentality of others, must amplify and coordinate the instrumental potential of his own body parts. These should cease to be independent pleasure receptors and instead take up their productive collaboration. This presumably universal process varies in its effects by sex, by culture, and by the individual's location on the folk-urban continuum of his society. Thus young women, particularly young mothers, generally rely on the instrumentality of men; by the same token, they are more diffusely sensual in their erotic makeup. Urbanized men also rely on instrumentalities and resources outside of themselves; they do not repress and compartmentalize their sensuality to the same degree as remote-dwelling men of the same ethnicity and culture.

Those people—urban dwellers; affluent, alienated youth; young mothers—who are to some extent buffered against the iron rule of necessity have had less opportunity and certainly less incentive to limit their various capacities for pleasure; consequently, in later life they may turn less vehemently toward sensual receptivity. But in the case of men honed early to the productive life, there appears to be a later-life relaxation of the impulse to produce, with a consequent return of the repressed receptive tendencies, in such a way that the individual moves from productivity to receptivity, and becomes, as in childhood, dependent on the Active Mastery and instrumentality of others. Consequently the developmental sequences of early life are reversed, and the capacity for erotic pleasure is no longer fenced off in the genitals: The eye can once again be absorbed in erotic looking, the mouth once again devoted to oral-erotic pleasures. Eros can ebb back to its original reservoirs of mouth, eyes, skin, and sphincter; it can be fractioned according to its original archaic, anarchistic goals.

This argument supports our major contention, that the psychology of later life is founded on substitution and even expansion rather than sheer depletion. To repeat, the inner sources of masculine aggression and power are depleted, but they do not leave a vacuum; rather, they are replaced by strivings toward goals and pleasures of a different sort, having to do with the somatic receptors—tongue, eyes, and skin—rather than the body's effectors of limb, muscle, and erectile tissue. In effect, the change is to a more narcissistic or omnivorous position of the psyche, in which an early demand is revived: that all pleasures, through all modes and all zones, be declared mutually compatible and be made equally available to the self.

Having begun by tracing the oral propensities of later life, I end this chapter with the recognition that later life is not characterized by mouth hungers alone, but by a more global, inclusive erotic position, in which all the modes, those maturely genital and those pregenital, are preserved. It is not that the appetite for genital sex gives way completely to pregenitality; rather, the exclusive concentration on "mature" genitality gives way to a wider, more inclusive erotic spectrum, in which previously incompatible strivings are more or less equally represented. It is not only orality that surges in later life, but a more inclusive narcissism, a thrust toward omnipotentiality and omnisatisfaction. This appears to be the protean psychological event in the later life of men and probably of women: New forces have been released, for better or for worse. They can lead to a new construction of self and to new services for society or, like any potent force, they can destroy their own vessel.

In the following chapters, I will examine the fate of women, specifying the engines that power the seminal developments of later life, as well as the reciprocal conditions that shunt women's surgent energies toward new, sculpted outcomes on the one hand and toward new forms of psychological vulnerability on the other.

6

THE INNER LIBERATION
OF THE OLDER WOMAN:
PSYCHOLOGICAL AND
FANTASY MEASURES

> At a certain age the men writers change into Old
> Mother Hubbard. The women writers become Joan
> of Arc without fighting.
> —ERNEST HEMINGWAY, 1899–1961
> (from *The Green Hills of Africa*)

Studies of aging women provide further evidence of a developmental ordering of the psychology of later life. If the aging psyche were no more than a register of depletions and reactions to depletions, then we should find men and women, who suffer from much the same losses of kin, opportunity, and somatic integrity, aging psychologically along parallel tracks. But the available psychological and ethnographic evidence gives us strong indication that women do not replicate but instead reverse the order of male aging: Whereas adult males start from a grounding in Active Mastery and move toward Passive Mastery, women are at first grounded in Passive Mastery, characterized by dependence on and even deference to the husband, but surge in later life toward Active Mastery, including autonomy from and even dominion over the husband. Across cultures, and with age, they seem to become more authoritative, more effective, and less willing to trade submission for security.

While the species-typical psychological changes in the male can possibly be read as side effects of or compensation for exhaustion and discouragement, the changes in older women, which tend toward the same Active Mastery position held by younger men, strongly suggest addition rather than loss. And, because the changes in women are complementary to and in phase with those observed in men, it is not likely that older women alone move on a developmental tide that ignores men. The major changes in men that are symmetrical to developmental changes in women are, at least inferentially, equally developmental in nature. As we turn to consider women, we may begin to glimpse the outlines of an overarching developmental event of later life that includes both sexes, that binds them together in a dynamic intergender process.

Studying the Surgent Phase of Women's Development

Earlier I proposed that any developmental advance consisted of a sequence of operations, directed toward different parts of the individual's internal and external worlds. During the surgent phase of development, these operations shake up the inner self, as emerging potentials move toward communication through the language of unconscious symbolism, the private utterance of dreams and ruminations. In the proactive phase, the unfolding qualities are displayed toward outsiders, an intimate and trusted audience of peers or mentors. It is only at the third or sculpted level that the new endowments, now neutralized and legitimized through their expression in standard roles and common idiom, are displayed publicly, to the world at large. We will follow this sequence as we consider the developmental progression of the older woman. In this chapter, I will review the ways in which surgent potentials are displayed through fantasy, imagery, art, and empirical tests of deep psychological functioning. In the subsequent chapter, I will describe the ways in which emerging endowments are displayed in an increasingly direct, overt form, within the semiprivate enclaves of the extended family and finally in the most formal, public roles. Thus, this chapter will consider the kinds of fantasy that projective stimuli elicit from older women or from men about older women, and the following chapter will present the ethnographic data that describe actual ascendant behavior on the part of older women in private and public settings.

The Family Card

I got my first glimpse of the "androgyny of later life"—the tendency of older men and women to resemble each other psychologically—in the course of doctoral research using data developed by a special TAT card (see fig. 6.1). Concerned with intergenerational and family relations, the Family card shows an older man and an older woman, as well as a younger man and a younger woman. Younger American subjects, men and women in the age range 40–54, tended to see the older man dealing in a commanding manner with some domestic problem that had been precipitated by the younger couple. Typically, in their eyes, the younger couple were asking for permission to get married or to seek a better job away from home; the older man defended their move toward maturity and independence against the unsuccessful but often strident opposition of the older woman. However, in the eyes of the older research subjects, aged 55 to 75, the problem before the family might remain the same, but the older woman had the dominant voice in resolving it. And while the older man might still back the young against the mother, he is no longer seen as effective—he cannot influence the outcome in their favor. In the eyes of older subjects, the older male figure became somewhat "maternal": concerned, even loving, but in terms of power, ineffectual. On the other hand, the older woman figure was seen to inherit the social power that the older man had lost, as well as his "masculine" physical qualities: "Look at that bull neck—she wears the pants, she runs the show." The indicated sex-role reversals are depicted in figures 6.2 and 6.3.

The Male Authority Card

In chapter 3 I discussed the age changes in the male perception of women, as captured by the Heterosexual Conflict card of the transcultural battery. In results that parallel those generated in Kansas City by the Family card, the Heterosexual Conflict card elicits perceptions of a powerful woman from the oldest male subjects, across the entire cultural range of my study.

Similar age trends are elicited even by TAT cards that do not

FIGURE 6.1
The Family Card

explicitly depict women. This same sex-role turnover is dramatically captured by the Male Authority card (see fig. 6.4). While Druze men below age 60 almost invariably see this stimulus as depicting men exclusively, a number of men over 60 see the older male figure as a beggar, asking for food or money from a woman, who may or may not indulge him. Again, the tendency to see a young male as a dominant woman, or to see a possibly authoritative old man as a beggar, is not a cohort phenomenon, limited to a particular generation of Druze men. Longitudinal studies with this card reveal that nine Druze, all but one of them over 60 years of age, who saw the older man as

FIGURE 6.2

The Family Card: Role Descriptions of the Older Woman Figure
(Kansas City Subjects)

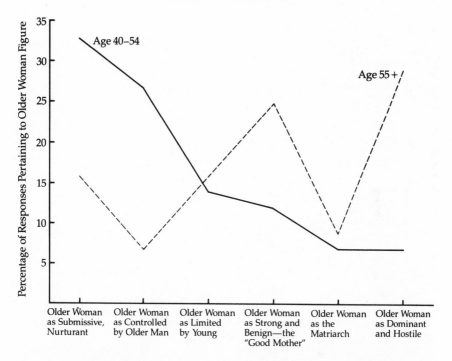

an authority toward other males at Time$_1$, saw him as begging from a woman at Time$_2$. This intraindividual shift is all the more striking in that Druze values favor fierce self-reliance, and their mores do not sanction begging on the part of any age group, least of all for the elderly guardians of the society's religions and traditions. Such imagery is the rendering of unconscious desire; this projected wish on the part of older Druze, to become passive and needful before a strong woman, cannot be traced to any general custom or usage of their community. Indeed, these covert wishes fly in the face of Druze custom (which is why they remain unconscious) and can be seen only as the signature of internal, self-initiated developments, which eventually bring about some pacification of the once-authoritative older man.

FIGURE 6.3

The Family Card: Role Descriptions of the Older Man Figure
(Kansas City Subjects)

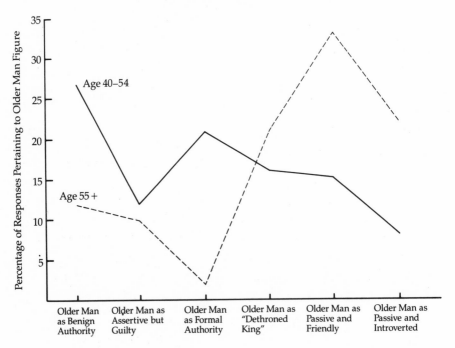

The Kansas City TAT Findings

The changing nature of the older woman is partly inferred from older men's respectful fantasies about women. These changes can also be inferred from the plots and organizational qualities of American women's stories elicited by a TAT battery containing a number of cards and representing a variety of psychological issues. In this instance, the respondent's stories were not analyzed one at a time, on a card-by-card basis; instead, each collection of stories, or protocol, was analyzed on a subject-by-subject basis, without knowledge of the respondent's age. Then, just as in the single-card analysis, the full protocols given by individual women were classified as to the major mastery style that they expressed, into the following categories:

FIGURE 6.4
The Male Authority Card

Active Mastery

The two subgroups included under this heading are the Rebellious Daughters and the Moralistic Matriarchs. Both groups of women justify effective, focused action upon the environment by finding enemies to fight and worthy causes to uphold.

For the Rebellious Daughters, the "enemy" is a restrictive (usually maternal) authority, and their identification is with young people who struggle for their autonomy and self-determination against such authorities. The following story, inspired by a card (the Farm/Family scene) that shows two women and a man in a bucolic setting, is typical:

> Well, this girl lives on a farm, and she hates it. Her mother is very defiant and wants her to stay. [Yes?] But she goes on and becomes a very good scholar and teacher. [Why does she hate the farm?] It just doesn't appeal to her. She likes books and knowledge better.

It may be that these women identify with the rebellious young in order to disavow their own infantilizing, authoritarian tendencies.

139

If so, they turn what could be a disquieting inner struggle into a brisk engagement with the world.

The Moralistic Matriarchs give equally free-swinging, although more idiosyncratic, stories. For this group, the opponents are callow, misinformed, morally lax young people. The role of the matriarch is to guide and discipline the errant children. A typical example is this story offered by a woman in response to a TAT card that, because it shows a younger and older man, is suggestive of father-son interaction:

> This one leaves me kind of cold. . . . The young man is certainly quite sullen looking. . . . The older man is quite kindly looking, and I can't connect the two as having any relation, as this other one has such an awful sour puss, so I can't think that it is his son especially. . . . You could say that this fellow is a dope addict—that is what he looks like—and an elderly gentleman is trying to straighten him out and make a man of him. About the future, it looks pretty black. . . . [Why?] He looks like he has such a weak character, you wouldn't expect him to come out of it.

As in the above example, the women of this subgroup justify vigorous, intrusive action through their struggle with bad children rather than with bad parental authorities. They are alert to the moral deficiencies of others and having found them, they take direct, corrective, and retaliatory action. Essentially, as the dope-addict story suggests, they rephrase personal struggles and everyday intergenerational frictions into battles between right and wrong, with themselves on the side of the right. Thus these women feel indignation over the moral deficiencies that they read into the stimulus figures, conveniently unaware that it was they themselves who put them there.

Despite their contrasting identifications, these two female subgroups are alike in their essential technique for sustaining inner balance and outward effectiveness: They maintain Active Mastery by externalizing their own guilty feelings. Their superegos are deployed not against the "bad-me" aspects of self, but against blameworthy, "not-me" aspects of the world. Thus freed from the nagging of a bad conscience, these women can perhaps act in an expansive, goal-directed fashion. As younger women, their self-esteem requires a bad parent, and as older women, it requires a bad child. Although they may sometimes stretch the facts in order to find a proper target, we can infer from their logically developed, well-organized stories that their actual behavior is reasonably effective and stays within conventional bounds.

Men and women of the Active Mastery type are alike in terms

of the formal, structural qualities of their stories. Both sexes impart vigor and conflict to scenes depicted on TAT cards in a framework of logically developed, well-organized, and reasonable stories that are generally positive in their outcomes. At the same time, there is a major sex difference. Active Mastery men tend to propose that their struggle is with a refractory but impersonal environment or task, and sometimes with those passive aspects of themselves that interfere with achievement. For the women, the conflict is between people, and issues are thrashed out in terms of their interpersonal connotations and complications. For the men, the interpersonal drama tends to be muted, and the focus is on the task and the hero's capacity to deal with it. In parallel stories told by the Active Mastery women, the interpersonal issues are the crux of the plot, and decisions are fought out openly, in a vivid, sharply delineated social arena.

Passive Mastery

In contrast to the lively, extrapunitive Active Mastery women, the Passive Mastery women are intropunitive—they turn their own aggression against themselves. They solve internal or external conflict through changing themselves rather than the world. Two subgroups are included: the Maternal Altruists, for whom the intropunitive style is effective, and the Passive Aggressors, who seem to lack the external supports that could validate and redeem the intropunitive style and for whom, therefore, the style is ineffective.

The Maternal Altruists are similar to the Rebellious Daughters of the Active Mastery type in that they share their maternal interests and their altruistic identification with children. However, while the Rebellious Daughters rely on externalization as a defensive technique, the women grouped here hide their anger under a smiling exterior. The Maternal Altruists are not looking for a fight, nor are they seeking to dominate some objectionable aspect of the external world in order to control troubling aspects of themselves. Rather, much like the passive, and pacified, older men, they are uncomfortable with aggression, and they look for placid, predictable milieus, benign surroundings that support them in the denial of their aggression and in their conscious identities as nurturing, maternal individuals.

The following story, told in response to a TAT card that suggests mother-son relations, illustrates the self-effacing qualities of this group; here, despite her reluctance and inner despair, the heroine puts the son's career above her own need for companionship. Rather than involve the son in her conflict, the mother turns to busywork. This emphasis on good works suggests that the respondent handles any unvoiced anger by imposing some penance on herself.

> The mother is sorry that the son is going away. She hates to see him leave, but she doesn't want to stand in his way. . . . [What is she thinking?] She is thinking of how lonely she will be but that she must not let him know. . . . [How does it end?] She'll get busy with club work and things, and lead a happy, useful life.

The Passive Aggressors, seemingly the more depressed members of the female Passive Mastery group, are represented by constricted women who, much like equivalent male groups, defend themselves against both external and internal stimulation, and conform in a rather apathetic manner to conventional guidelines for behavior. They stress depressive story themes, but otherwise do not go much beyond a sparse, although accurate, description of the card features. The following story, told in response to the Farm/Family TAT card, suggests the somewhat inert and self-pitying quality of this group:

> I don't know. . . . That's her daughter, isn't it? . . . [Up to you.] She looks like her. Her son is plowing. She's come from school. They are sad, just like me. . . . [How does it end?] Sometimes things turn out good, sometimes bad. We just have to take the heartaches.

Unlike Active Mastery women, the passive women see themselves as victims rather than as aggressive masters of their fate. Although they may employ some of the moralistic, self-justifying tactics of the more assertive women, their use of externalization does not liberate their aggressive energies. These passive individuals cannot go into action, even when they have a worthy cause.

In some cases, the depressed mood of these women represents a socialized rephrasing of their fundamental anger. They resent very strongly the departure of their children, but they cannot act on their retentive wishes. They cannot resign themselves to their fate, yet they cannot fight it. Their coping techniques are essentially passive and self-directed—they overconform to external demands; their obedience gives them the right to expect restitution and support, or to feel aggrieved if they are not rewarded. Thus these subjects are indirectly manipulative rather than directly intrusive. They exert influence primarily through the constraints and demands that they impose upon themselves and through the guilt that their deprived, depressed condition arouses in others.

In terms of both the form and content of their fantasy, the Passive Mastery women are somewhat more dramatic than their male counterparts. The women announce their depression (and thus partly transcend it), while the men seem to live it out. However despairing they may feel, these women continue to communicate and to externalize their inner states; their depression serves as the currency and idiom of their traffic with others. The men withdraw more quickly from interpersonal communication to private rumination, internal dialogue with themselves.

Magical Mastery

The stories told by the Magical Mastery women are notable for the open and often inappropriate display of primitive, aggressive, retentive, and sexual feelings. Such motivations are often ascribed to the stimulus figures, with little regard for the objective, realistic card features. These women grossly confound their reactions to the cards with their perceptions of them. Their ego boundaries seem unclear; the world is an unrecognized cognate of themselves, its nature and reality defined by their feelings toward it. In effect, they confuse the messenger with the bad news.

The following story, told in response to the Rope Climber card, illustrates how this group maintains emotional equilibrium through moralistic and zestful condemnation of their own unrecognized and projected sexuality and aggression:

> This is evolution for you—back to the cave man, back to the monkeys. I have read about Neander's man. . . . This sure looks like the cave-man age—the taming of the shrew—the men back then were small in stature. Look at his muscles. . . . [What is he thinking?] He thinks only of the carnal—kill, protect himself, get food. . . . [How does it end?] I don't believe that time will ever come again, do you? I read in *Quick Magazine* that in one thousand years there will be no race segregation. The black race will supplant the white. . . . That's what comes of Negroes and whites going to school together. When love strikes, they intermarry.

In the following story, told in response to the Heterosexual Conflict card, the man's sinfulness and the woman's holiness are taken for granted. Almost explicitly, the woman is an agent of God, and, thus justified, the respondent's autocratic tendencies clearly break through:

> I think this little lady is trying to tell him about the Way, but he is just

objecting, turning himself away. . . . [What are they thinking?] Well, I don't know if he has a need in his life and she's trying to help. He needs the Lord. . . . help from his Heavenly Father. . . . [How does it end?] She'll win out. She is going to pray for him.

Thus women in the Magical Mastery group grossly and often rather blandly distort and misperceive reality in order to justify impulsive and domineering behavior. They experience little internal conflict concerning their primitive needs. Their fight is with the environment, not with themselves. In a sense, these women are rather like the nearsighted Mr. Magoo—caricatures of inner direction. They see the world as they wish to see it, and they may consequently act with such self-confidence that others grudgingly confirm and conform to their idiosyncratic versions of reality.

The proportion of women in the two nurturant subtypes (Rebellious Daughters and Maternal Altruists) drops off markedly with increasing age: 41 percent of the 40- to 49-year-old women are found under these headings, whereas only 38 percent of women in their fifties and 18 percent of the women in the oldest group are located here. The proportion of women in the self-centered categories (Passive Aggressors and Magical Mastery) shows a corresponding increase with age (see fig. 6.5).

Thus the younger women demonstrate a nurturant, positive attitude toward the young, including an acceptance and espousal of their autonomy. Older women offer a more retentive, self-centered version of mothering, they visualize direct conflict between themselves and their children, and they are not averse to self-delusion in order to justify making self-serving claims on their offspring.

Perhaps the attenuation of the maternal role and the growing independence of the young allow some of these older women to take a more overtly self-centered, managerial position. In effect, older women might permit themselves the self-indulgence that had to be denied when they were primarily concerned with the rearing of children. Just as men in later life seem to turn back upon themselves the outward-directed aggression of their youth, so older women may turn back upon themselves the nurturance that no longer finds its appropriate external target. They may become their own lost child and claim for themselves the indulgence that the younger women still claim and fight for as the right of their children. In addition, knowing that their children are now strong enough to stand against them, they may permit themselves a more autocratic, intrusive role in regard to the

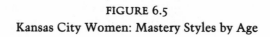

FIGURE 6.5

Kansas City Women: Mastery Styles by Age

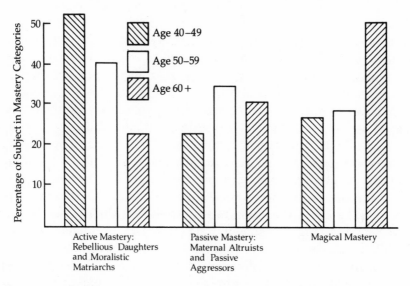

younger generation. In effect, knowing that their assertiveness will no longer damage or stunt their kids, they can begin to enjoy it.

The Mastery Types and Female Satisfaction

When the mastery types for women were studied in relation to other, validating measures, such as the Life Satisfaction rating developed by Bernice Neugarten, Robert Havighurst, and Sheldon Tobin (1968), the results did not accord with predictions based on the male sample. For men, the sense of well-being as measured by the Life Satisfaction Scale is linked to the level of psychological functioning and seems to decline in step with the shift away from Active Mastery to the more primitive state, Magical Mastery. But in the female case, the Magical Mastery women, those who approached the TAT in arbitrary and distorting ways, often attained the highest Life Satisfaction ratings.

This unpredicted finding suggested that the mastery typology should be differently interpreted for women and men. Reassessment of the data showed that psychological parameters independent of the usual criteria for good ego functioning might account for much of the variance in the female data. Specifically, inward-turned, though realistic, Passive Mastery women seemed to receive lower Life Satisfaction ratings than did the more externalizing, less realistic Magical Mastery women. Thus the Life Satisfaction ratings seemed to distinguish between outward-turned and inward-turned styles in women, but not between realistic and unrealistic ego functioning.

Accordingly, the female groups were reordered. The Passive Mastery women were rebaptized as intropunitive, while both the Active and Magical Mastery groups, composed of those who turn aggression outward, were lumped together under the extrapunitive heading. The Life Satisfaction ratings were then distributed against these new groupings. As shown in figure 6.6, the difference between mean Life Satisfaction scores was now in the predicted direction. A statistically significant number of extrapunitive women, including many of the Magical Mastery group, achieved Life Satisfaction scores well above the median, as compared to the more reality-oriented but intropunitive women. Thus, viewed against the psychoanalytic models of mental health, the extrapunitive women were often inferior to the intropunitives, but no such ego "weakness" shows up in their Life Satisfaction ratings. This finding suggests that contentment for women, by contrast to men, might be related more to the deployment of their aggressive energies than to their capacities for strict reality testing. A vigorous, intense engagement with the world, even when such activity is rationalized by unrealistic ideas, may sponsor greater happiness for American women than a more controlled stance, however reasonable the latter might be.

The younger Active Mastery women fight well-rationalized battles against intrusive parents or rebellious young; they enforce discipline and demand respect, or conversely, fight for the right of the young to gain independence. The older Magical Mastery women seem to determinedly distort reality in order to establish grounds for what may be their continued autocratic participation in the lives of their now-independent children and their somewhat emotionally withdrawn husbands. Essentially, despite important differences in reality-testing, both the older and the younger extrapunitive women look for a cause and an enemy; it is the psychological tactics involved in maintaining this belligerent orientation that changes over the years. Thus

FIGURE 6.6

Distribution of Extropunitive and Intropunitive
Kansas City Women by Life Satisfaction Ratings (LSR)

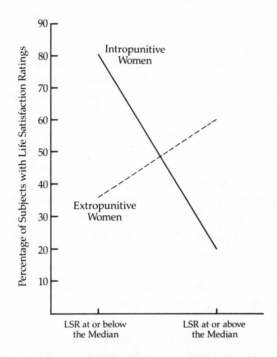

younger American women—who have vulnerable children, as well as (sometimes) intrusive mothers, mothers-in-law, and even assaultive husbands—do not have to distort their social reality in order to find a cause or an enemy; but older women can be exempt from the battle, unless they choose to enter it. To do so, they may have to turn relatively inoffensive kinfolk into obstacles or opponents. They bring this transformation about internally, by altering their perception of others without too much obsessive regard for external realities. A late-life reliance on Magical Mastery allows them to conserve the active stances of earlier years, outside of the family, as well as within it.

Age and Gender Differences:
American and Western Data

A *potpourri* of psychological studies into the more surgent aspects of American and other Western female personality supports the general notion of a more extroverted and free-swinging "second life" for older women.

Thus Brooks Brenneis (1975), who contrasted the dreams of women aged 40 to 85 to those of younger women, was struck by the "augmented amount of internally represented robust and locomotor activity for the older women." The dreams of younger women do not show them as being physically vigorous to the same degree (though they do dramatize much greater concern about being either the source or the victim of aggression). Along lines that I will suggest in chapter 8, Brenneis speculates that, as older women move away from parenthood, they cease to worry about the destructive consequences for children of either their own or others' aggression; by the same token, they feel freed up for vigorous, outgoing activity.

The shift toward a more active posture for the older woman is registered in daydreams as well as night dreams. Thus Giambra (1973) found younger men and women, up to the age of 40, to be about equal in their production of achievement-centered daydreams; however, particularly after age 70, there is a decisive shift toward the predominance of such themes in the daydreams of older women. (This could, of course, be a sampling rather than a development effect: The vigorous post-70 survivors could also be the most achieving women.)

Further evidence for structural age changes that have cognitive as well as emotional consequences comes from psychological testing of the Jewish population of Israel. The investigations of Shanan (1978) into age and ethnic differences in the Israeli population led to the finding that, in masculine protocols, "futurity" dropped off with increasing age. Thus, as they grew older, Israeli men, regardless of ethnic subgroup, were less likely to propose TAT story outcomes that referred to a future beyond the immediate situation. But this was not the case with women, for whom futurity increased with age, again, regardless of ethnic or educational background. The capacity to "construct" and give reality to the future has generally been regarded as a masculine trait; yet in these findings, we see women taking on, with advancing age, the ego modalities that were thought to be concentrated in men.

Studies pitched at the more intrapsychic level also turn up "male" characteristics in an older female population. Thus E. Lowell Kelly (1955) reported a trend toward increasing masculinity in women, based on a longitudinal study covering the age range from the twenties through the forties. His results are confirmed by Paul Cameron (1967).

Magical Mastery in the Service of Assertive Action: Cross-Cultural Indications

This compounding of Active and Magical Mastery in support of the older woman's feisty, combative stance is not limited to the American scene but seems to have planetary distribution. For example, Kakusho Tachibana (1962) tested Japanese men and women on an introversion-extroversion scale, and found that men are more extroversive than women until age sixty, when a decline on this variable puts their scores below those of women. By contrast, the female extroversion profile remains fairly constant, rising slightly across successive age groups.

Though based on cross-sectional data, Tachibana's findings can nevertheless be read longitudinally: There is no good reason to believe that older Japanese women, who received their early socialization during an era that valued female modesty, have been more sexually extroverted than their younger sisters across their entire lives. Since extroversion implies a greater interest in the nature and management of external affairs, its later-life appearance in Japanese women may be part of the same developmental transition, toward outgoing assertiveness, that we have already found in the TAT protocols from older women in the United States.

Marcello Cesa-Bianchi (1962) also found that, in Italian nursing homes, older women held their edge over older men as regards vigor and the free expression of anger. He observed male and female residents of a Milanese old-age home, aged seventy and over; in that difficult setting he found that women worked more than men, they showed higher intellectual efficiency than men, and they were more self-directing than men. Consistent with their tendency to push ac-

149

tively into the world, the older women scored much higher on an index of extrapunitiveness, implying a greater tendency to locate the blame for their troubles outside of themselves. Cesa-Bianchi's findings converge with Tachibana's from Japan, prompting the notion that extroversion and extrapunitiveness are linked in older women (just as introversion and intropunitiveness appear to be linked in older men), with the first tendency facilitating the second. That is, the extroverted woman can discover the source of her troubles (hence, occasions for blame) outside of herself. Freed from the nagging sense of guilt and responsibility, the extroverted woman can happily turn aggression outward against others to become their bad conscience instead of her own.

Female projective data from the Druze society adds to a major finding from the Kansas City data: that the more open expression of women's aggression is sponsored by idiosyncratic thinking. The older Druze women show, quite markedly, the increase in Magical Mastery that we have already traced in the American materials. Across all cards of the Druze battery, the proportion of idiosyncratic, "magical" responses given by older women (37 percent) is significantly higher than the proportion of such responses given by older Druze men (16 percent). Furthermore, although the percentage of "magical" responses goes up significantly with age for both sexes, the female proportion for such responses is consistently higher, at any age, than that for Druze men.[1]

Most notable is the older Druze women's resort to unrealistic (by Druze response norms) perceptions of damage, death, "ugliness," and illness: "These have blind, ugly eyes," "This is a dead person," "This is an owl," "These are beggars," etc.

Incidentally, these rather striking sex differences point up the dynamic (rather than the organic) bases of such later-life card distortions, for both men and women. Clearly, there is no reason to predict a greater incidence of presenile brain syndromes in older women than in older men of the same society. It follows that the higher incidence among aged females of qualitatively poor responses points to the voluntary (though unconscious) reliance on primitive defenses—denial and projection—rather than to the involuntary consequences of brain damage in this group. This dynamic interpretation is reinforced when we consider the marked variations, by cards, in the distribution of the older women's inappropriate, magical responses. Thus cards that portray vigorous younger men (the Rope Climber, Male Authority, Heterosexual Conflict, and Horse and Men scenes) elicit a much lower mean percentage of inappropriate responses from older women than do cards that portray women, ambiguous figures, or old people. Thus the "masculine" cards almost completely eradicate the male-female gap in regard to such "magical" responses. Where the older male mean percentage of magical responses to the "masculine" cards is 15 percent, for older women it is only slightly higher, at 18 percent. By contrast, the remaining cards, with less clearly "masculine" latent stimulus demand, elicit much sharper sex differences, the older female mean percentage for magical responses being 56 percent, against a male mean of 26 percent. Clearly, the "masculine" cards inhibit magical responses much more dramatically in the case of older women than in the case of older men; this interaction of sex, age, and card type in the production of "magical" responses again points up the dynamic rather than the purely organic significance of such distortions. If organic factors were involved in a major way, we would expect the degree of distortion to vary according to structural features of the cards—that is, their ambiguity—and not according to their content. This "cohering" effect of the "masculine" cards for older Druze women prepares us for a social phenomenon of later life that I will examine at length in the next chapter. This is the older woman's alliance with her grown son, against her daughter-in-law and also against her aging husband. These implicit mother-son partnerships—much like the content of the "masculine" TAT cards—appear to have a vitalizing effect on the senior woman.

Georgia O'Keeffe: The Escape From Interiority

Across cultures, depth psychological measures are unanimous in demonstrating the surgent virility of the older woman. This aspect shows up in spontaneous creative acts, such as works of art, as well as in the responses stimulated by psychological instruments.

As we have already seen, the lifework of male artists testifies to the inner psychological developments that they share with their less artistic fellow men. Thus far, I have not considered the life cycle of women painters, in part because there are not many who have attained the eminence that merits printed collections and retrospective exhibits. However, recently deceased after a long life, Georgia O'Keeffe

(1887–1986) was a contemporary female artist of much repute. Her canon gives graphic evidence of the changing inner life of the older woman. As we would predict, O'Keeffe reverses the usual masculine schedules regarding the depiction of space: Where the older male artists burrow into feminine "inner space," leaving the large perimeter behind, O'Keeffe ventures the other way, from inner murkiness into an open world charged with dry light. Figure 6.7 shows that, in her early years (from ages 20 through 45), O'Keeffe tended to paint a "tight-focus" panel of magnified objects: close-ups of flowers, shells, and bones—a mouse-eye view of the world. Forty-six percent of the paintings available to me from the early years are of this tight-focus variety.

Prototypical for this early period is her "flowers" theme: Twenty-seven percent of the early sample shows the interior world of invaginated, richly colored, sensual blossoms. But there are no such "flower" paintings after age 55, and only 6 percent of the later sample feature the close investigation of such small, soft objects. In old age, O'Keeffe's paintings are of vast spaces, arid desert regions, unpeopled but full of light. With age, she moved from the sexual, richly furnished, interior microworld, into a vast domain of dry sands, sharp edges, and roads that knife through the high plateaus. These are the lands of the perimeter, the proper settings for wide-ranging "masculine" action. Seventy-five percent of the later paintings are of this spacious and hard-edged variety, in which sensuality—to the degree that it is still present—belongs to rounded rock spires and sand dunes rather than to the soft inner flesh of flowers. O'Keeffe's reversal of the typical masculine sequence concerns the nature of line, as well as painterly technique. Whereas older male artists typically undo the hard-edged line of their youthful paintings, O'Keeffe moves the other way: from the diffuse, pulsing color of her youthful flowers to the sharp contours of mesa and arrowing road in her later years.

The View from Depth Psychology: Differences Between Older Men and Women

The later work of Georgia O'Keeffe captures the essence of the healthy, capable older woman: adventurous, expansive, self-asserting. By contrast, older men increasingly stress their self-control, their friendly

FIGURE 6.7

The Paintings of Georgia O'Keeffe: The Depiction of Space and Form

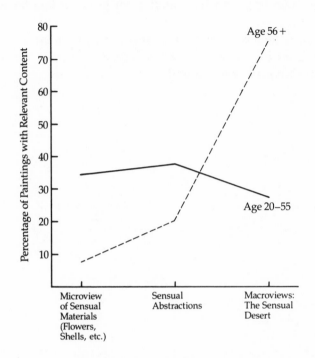

adaptability, and even their passivity. Action in the external world can give way to an anxious, compulsive reordering of the self. Where older men seek to control their spontaneous urges, older women appear to seek out opportunities for vigorous action and lively interpersonal encounters.

In terms of the nature and disposition of the ego defenses, men defend the external world against what they imagine to be the consequences of their destructive impulses, while women seem to defend their personal claims against the counterpressures of the world. That is, older men strive for quiescence and comfort. Their psychological defenses do not facilitate overt action but replace it with inhibitions. The substitution of rumination and fantasy for direct action may safeguard the desired passivity. The outer world is changed subjectively, through altered perception (Magical Mastery) instead of instrumental action, which has been renounced. By contrast, the "magical" defenses of older women may run interference for their unflagging efforts at direct control of the external world. Older women do not conjure up a dream world in which the goals of action have been magically achieved; rather, they confabulate versions of reality that permit them to act without anxiety or guilt, in flamboyant and dominant ways.

In the next chapter, I will trace the ways in which the surgent energies studied in this chapter drive a wide range of "standard," public role behaviors, across a wide range of societies.

7

THE VIRILE OLDER WOMAN: ACROSS HISTORY AND ACROSS CULTURES

> In the 18 years between Grandfather George's death and my arrival in Morrisonville, Ida Rebecca established herself as the iron ruler of a sprawling family empire. Her multitude of sons, some of them graying and middle-aged, were celebrated for miles around as good boys who listened to their mother. If one of them kicked over the traces, there was hell to pay until he fell obediently back into line. In Morrisonville everybody said, "It's her way or no way."
>
> Her sons' wives accepted the supremacy of mother-in-law rule as the price of peace and kept their resentments to themselves. When her boys married the women she approved, their wives were expected to surrender their swords in return for being allowed to keep their husbands for the spring planting.
>
> —RUSSELL BAKER, 1925–
> (from *Growing Up*)

In a previous chapter I described the male drift toward Passive Mastery that, in many societies, is written into the text of the age-grade systems that regulate the role behavior of older men. But social shaping of this sort is less evident in the later life of women. As noted in the chapter on the social regulation of males across the life cycle, age-graded controls on men appear to be strictest in warrior societies, acting to blunt the aggression of young men within the community, to keep it under the braking hand of old men and to shunt it outward, against enemy, prey, and unconquered nature. Perhaps women, possessing more capacity for eros than thanatos, are innately more trustworthy than men: Their kinder nature will lead them into the right and preserving paths without too much coercion by social nurture. For whatever reasons, the surgent, ascendant powers of older women as picked up by our depth-psychological instruments are not registered in the formal texts and codes of age-graded societies. In the usually patriarchal folk society, the older woman's move toward matriarchy is, in the formal sense, a nonevent. While most informants are aware of it, this powerful development is given little institutional expression or recognition. Nevertheless, the unofficial matriarchy of later life is so apparent that it has been recorded by many independent anthropological observers: Thus Susan Gold (1960) asked twenty-six ethnographers, varied as to their theoretical interests, to report on any age-related changes in sex role in the varied groups that they had studied. Fourteen reported a shift toward greater female dominance in later life. The remaining twelve reported no change. In no case was the balance seen to swing, with advancing age, toward greater male authority over the wife.

The Age Grading of Female Dominance

Empirical and anecdotal evidence from various sectors of American society bear out this transcultural pattern. Consider, for example, the findings from a landmark study of life-cycle transitions in the United States. The findings of Marjorie Lowenthal, Majda Thurnher, and David Chiriboga (1975) concerning middle-aged urban American couples could serve as a model for many of the cultures that we will review in this chapter. Briefly, these authors find that middle-aged, postparental American women become more dominant in the family and take a more maternal but also more managerial role toward the husband. The husbands do not look for new arenas to master but instead desire self-indulgence and release from responsibility. In effect, they move into the filial niche that has been vacated by their grown children. Even as older men look to interpersonal rather than inner resources for their security, this same people-centeredness becomes less important for their aging wives.

Lowenthal's data suggest that this female entry into a more independent existence is often preceded, in early middle age, by a period of extreme malaise, during which time women can suffer greatly but inarticulately, with little hope for the future. Understandably, marital problems increase during this troubled time. But these midlife women retreat only in order to better advance, and after their late fifties they are less likely to report psychic symptoms or social troubles. The ebbing away of symptoms coincides with a greater readiness on the part of women to declare that they are the "boss in the family," that they are more assertive, that they are interested in work outside the home, and that they are more "masculine" than formerly. Their pain eases, and they can recognize and take pleasure in new-found "masculinity." As with other births, the woman's rebirth into a more omnipotential position may be preceded by some necessary pain.

In the American case, the social scientists are only confirming a folk knowledge that declared itself in popular media, even during more patriarchal periods. Some of the most popular comic strips of the early 1900s—"Moon Mullins," "The Katzenjammer Kids" (also known as "The Captain and the Kids"), "Bringing up Father" (or, "Maggie and Jiggs"), and "Our Boarding House"—featured powerful older women who routinely abused their aging husbands with rolling pins, in retaliation for the men's participation in forbidden poker games (as with the "boys," at Dinty Moore's).

But the legendary power of older women in the United States has never been restricted to the comics. Even in the presumably male-dominated precincts of Protestant churches, young priests have always been terrorized by aging deaconesses, deferring to them publicly, while maligning them privately as "dragons." In America the church has traditionally been seen as the province of women and sissified men: Little wonder that the dragons would rule these roosts. But in the later years of this century, women of all ages have moved, by multitudes, into domains that were previously exclusive to men, the professions where the serious work, the world's work, is done. Despite the fact that they were socialized during an epoch when female careerism was not encouraged, postparental (or widowed) American women routinely start out on careers. If they are already employed in the helping professions, delivering hands-on care, they typically move into administration, or into fields that promise money and power, rather than opportunities for service. Today midlife husbands

157

and wives cross paths as they shift professions: The aging husbands are leaving executive country to become social workers, and their wives are leaving the social work agency to attend law school.

And if older American women opt for new careers, they also stick to them longer than their husbands. Thus Robert Atchley (1976) finds that lack of a work role does not affect the morale of retired men as much as it affects the morale of their wives. (Philip Jaslow [1976] reports similar findings.) Atchley asks, "Can it be, then, contrary to the extant literature on adjustment in retirement, that work is actually a more significant factor in the lives of older women than men? Are the non-work roles available to older women actually less fulfilling or rewarding than is leisure for men?"

Alan Kerchoff (1966) reached similar conclusions as he studied morale and role change in postretirement couples. His data indicate "a greater sensitivity to interpersonal relations in the conjugal unit on the part of the husband and a greater concern for practical activities on the part of the wife."

Atchley's findings lead him to conclusions that parallel mine: "It would indicate a kind of role reversal from the presumed model husband and wife relationship in our society, which calls for the husband to emphasize an instrumental orientation, and for the wife to have more of an expressive orientation."

The same liveliness extends beyond the young-old to include the old-old American woman, as well. Robert Peterson (1979) tells us that in a community-dwelling group of urban aged, "the women especially seem to have a tenacious hold on life, and continue to survive in spite of chronic ill health." The same spirit animates the institutionalized old females, who, as reported by Paulig and McGee (1977), held paid as well as unpaid formal leadership positions in the two senior citizen centers that they studied; moreover, the most influential outside volunteers were also female.

The studies in the United States reveal the protean energy of the older women as it declares itself in a society that provides relatively few conventional roles for its organized expression. Despite this lack, older American women are not satisfied to put their energy behind rolling pins; we see their surgent vitality cutting new channels in whatever settings they find themselves: the family home, the world of commerce, and even the nursing home. The American data show us the role-creating, precedent-setting aspect of the older woman's vitality.' In the following sections we will learn how senior women have used their energies in more traditional societies, those longer-established than our own. This review will tell us something about those women's roles that have real staying power and, by implication, survival value for the women and their communities.

For the rest of this chapter, I will rely on transcultural data to illustrate the major routes toward power that older women have traveled and perhaps pioneered in various societies, and the nature of the powers that they have attained. I will spell out eight such syndromes, distinctive patterns of female striving and power; for each, I will cite the relevant ethnographic data and list all citations, by culture and region, that illustrate the particular motif.

The Older Woman as Ritual Leader

The first major, socially recurrent theme has to do with the postmenopausal acquisition, by older women, of the ritual powers and religious statuses that were previously closed to them. The ending of her procreative powers makes the older woman acceptable in sacred places and rituals on two counts. She is less sexually attractive, therefore less likely to stir men to lustful thoughts when their minds should be on God; and she is no longer a danger to the ritual—she will not pollute the service with her menstrual blood. Accordingly, after the menopause, the older woman can join in the circle of religious dancers or take on ritual tasks that are forbidden to fertile women. This motif is probably very common and may be synergistic with other patterns identified below. We have direct reports from the following groups:

Jack Goody (1962) observes that only the postmenopausal women of the LoDagaa (northwest Ghana) are allowed to wash corpses. He elaborates:

> It is prescribed that these women should have passed the menopause, that they should, to use the LoDagaa phrase, have "turned into men" (*lieba daba*). They can no longer perform the main task of women, the bearing of children, and are, as it were, asexual. This attribute permits them to carry out intimate physical acts on members of both sexes. . . . But there is a more general aspect of this role of the older women, and one with parallels in many other societies. In a social system in which authority is largely vested in males, there is a strong tendency to equate the authority figures with maleness, or at least with asexuality. . . . Among the LoDagaa, it works out the following way. Authority, ritual and otherwise, is normally vested in men; within the general category of women it is those past the menopause who most nearly approach the male.

In the Amerindian case, women do not necessarily take over the secular activities that men have abandoned in favor of religious pursuits; rather, they go where the older men are and join them in the sacred dances before the gods. Particularly, older Amerindian women express their liberation from menses through entry into forms of ritual and shamanistic practice that were previously off-limits to them.

Thus Pamela Amoss (1981), who studied the Coastal Salish of northwestern Canada, wrote:

> Women past child-bearing age could no longer pollute hunters or their gear, nor would they contaminate the berry patches or offend the salmon. . . . With the raising of these restrictions, new avenues of spiritual power opened up to both men and women. The old were often caretakers for people in dangerous liminal states—successful spirit questers, girls at menarche, women in childbirth, warriors returned from battle, mourners, and the recently dead. A grandmother was an ideal attendant for a girl at her first menstruation, because she was not only wise, experienced, and concerned about her grand-daughter but also impervious to the girl's sacred contagion.

Likewise Jerrold Levy (1967) notes that, among many Southwest Indian tribes, the postmenopausal woman acquired greater sexual freedom, as well as the right to participate in rituals and healing practices previously barred to them. Kardiner and Linton (1945) reported that, among the warrior tribes of the Southwest, such matrons were called "manly-hearted women," and with that title they acquired the ritual prerogatives of men.

Besides exercising considerable domestic power, traditional Iroquois matrons of the longhouse could hold religious and political office: They too were "manly-hearted." Among the North Piegan of the Canadian Blackfoot tribe, only postmenopausal (and wealthy) women could be "manly-hearted"; a young Piegan woman who claimed the same title would have invited censure for her unseemly boldness and ambition. At the northern reaches of the Amerindian range, Charles Hughes (1960) finds that the oldest clan member among the St. Lawrence Island Eskimos, even if it is a woman, will lead the group and make the vital decisions. By the same token, the old people are feared for their witchcraft powers, and both old men and old women can be practicing shamans.

The Amerindian pattern holds across continents. Napoleon Chagnon (1968) finds that the Yanomamo of the Amazon Basic are paradigmatic male chauvinists who practice female infanticide and use young women as pawns in the political games of male kinsmen. Moreover, wife-beating is a major male pastime, as is warfare, which has as its object the abduction of women from neighboring tribes. Yet, old women are relatively exempt from forms of male exploitation that afflict their younger sisters. They are treated kindly by their grown children and are among the most mobile members of Yanomamo society; they alone have the ritual protection that allows them to travel freely between hostile villages, on sorcery-infected jungle trails, to retrieve the bones of the slain. They alone escape the restrictions, burdening all men and most younger women, that are imposed by intertribal warfare, witchcraft, and politics.

The Sexy Older Woman

Power has two aspects: It is both "good" and beneficial, as well as "bad" and destructive. Thus, as the older woman loses the "bad" power of her menstrual blood, she can, in certain societies, become a vessel of good power, invulnerable to the bad power of others and able to help those—for example, younger, still menstruating women— who are in a spiritually vulnerable condition. However, in another sense, once quit of their menstrual shame, older women can become "shameless," playing out immodest and even obscene behaviors.

Consistent with this "shame-free" motif, sexual lustiness, not necessarily woven into specific roles, has been noted among older women of several non-Western societies. Thus Clinton Grattan (1948) describes the privileged position of the older Samoan woman: "An old woman, or especially a group of old women, may say things that no one else would dare to say."

Similarly, Judith Brown (1985), who has also extensively reviewed the cross-cultural literature on older women, reports: "In many societies various improprieties are overlooked in older women: Older women may urinate in public, violate food and language taboos, or even drink too much on ceremonial occasions."

Robert Smith (1961) concurs, observing that Asian women in Korea, China, and Japan "find that the later years of life afford them the first small measure of freedom that they have known since early childhood." There is, in effect, a "greening" of older Japanese women: Those over age sixty can do lewd dances and make sexual jokes in mixed company—behaviors completely out of bounds for younger women. The same motif, of postmenopausal sexual liberation, permeates many ethnographic accounts from the Middle East. Thus, students of Lebanese village life have noted that postmenopausal women become quite bawdy, aggressive, and even controlling vis-à-vis males

161

of any age. In short, the bawdy older woman becomes like one of the boys, cackling obscenely with her buddies.

These last observations are particularly significant, in that they seem to violate cherished Muslim beliefs concerning the proper roles of women. Whether in the Middle East or elsewhere, the honor of the Muslim man is typically in pawn to his wife, for despite her subjugation she can dishonor him, and in a real sense emasculate him, through sexual adventures with other men. Accordingly, the extra-domestic life of women is usually severely restricted, to prevent any occasion for adultery. However, this strictest of taboos is apt to lift in later life, over all Islamic settings. Thus, in the Kelantan Malay villages, younger Muslim women are required to be modest in dress and behavior, and publicly deferential toward their husbands. But while younger women are not even permitted to visit coffee houses, older women have been known to run them.

The Self-Sufficient Older Woman

The lifting of menstrual taboos brings about social and sexual freedoms that younger women did not enjoy. Aging also reveals an advantage, a capacity for self-maintenance, that women have always had and that becomes most evident when they are widowed.

Colson and Scudder (1981) make this point with material from the African Tonga:

> An elderly widow, despite her poverty, is less handicapped than an old man without a wife. She gets along so long as she can grow and cook her own food, and can borrow a child to carry wood and water and stamp grain. Her demands on kin for labor to build a house or clear a field are sporadic and if not met at once cause her no great inconvenience. But an old man needs daily services that only a housewife will provide.

Regarding the Asmat of New Guinea, Peter Van Arsdale (1981) makes a similar point. Older women publicly degrade their husbands, he reports, but they can afford to, because they have less need of them. While the older Asmat need women, whether their own wives or the wives of other men, to cook for them, the old widow can grow and cook her own food. She needs only the occasional assistance of a child.

Among the Chipewyan Indians of the Canadian northwest, female versions of competence become, in later life, the standard for both sexes, and the aging man is thereby disadvantaged. According to Harry

Sharp (1981), competence is greatly valued by the Chipewyan, and the end of competence marks the beginning of old age and dying. The Chipewyan male can be truly competent only in the hunt—an area that requires much strength and endurance—whereas the older woman can retain competence in significant domestic activities that do not require the strength and endurance of a young body. Accordingly, the aging Chipewyan men lose their only claim to prestige, while their still vital, still competent wives retain dominance in the domestic zone, the last remaining sphere in which it can be manifested.

The Older Woman as Family Matriarch

While the menopause seems to be one releaser of latent feminine powers, equivalent senescent changes in the husband may be another. Thus a major motif of the later years concerns the older woman's relationship, not to her own body but to her aging husband. In this theme, the aging husband gives up interest in secular power and in the management of the home. Quit of his mundane powers, he becomes religious and, in compensation, links himself through ritual means to the power of the gods. The aging wife, often in company with the oldest son, then moves into the socket of power and dominion that the aging husband has abandoned. Presented below are some illustrations of this theme.

Kracke (1977) provides an example, from Amazonian Brazil, of the older man becoming a candidate for supernatural power, while his aging wife takes over the secular prerogatives that he has put aside:

> Although very different from one another, the older Kagwahiv I have known do seem to me to verify generalizations that have been made about the tendency of old men to revert to a more passive, dependent adaptation, and to turn to inward spiritual and religious interests, while old women become more active and dominating.

Sharp (1981) reports that the phasing out of parenthood brings about the predicted role reversals: The Chipewyan man dominates his wife until after the children are raised, at which time the statuses reverse. According to Sharp, male dominance pivots on two points: the social advantage granted by community structure and the superior physical strength that allows him to exploit his leverage. However,

> As a man's strength begins to fail he loses the ability to exercise his influence, leaving his position with respect to his children weaker than his wife's. At the same time, his loss of strength allows his wife to escape

163

the physical domination he has exercised over her and indeed often allows the wife to become physically dominant over her husband.

Margaret Mead's remarks (1928) concerning the powerful position of the older Samoan woman within the family capture the essence of a theme that has planetary distribution: "These older women are usually more a power within the family than the old men. The old men rule partly by the authority conferred by their titles, but their wives and sisters rule by force of personality and knowledge of human nature."

Mead's observation is particularly well illustrated by the studies of Peter Van Arsdale (1981) among the Asmat of New Guinea. According to him, the public sector of precontact Asmat society was dominated by powerful, titled elderly men, the *Tesmaypits*. However (in accord with Mead's dictum), these male gerontocrats deferred in the home to the powerful personalities of their aging wives. As in other societies, this division of authority was a later-life development. According to Van Arsdale:

> In the precontact era, as well as today, a woman was expected to be a subservient, efficient, docile mate. Younger women who did not measure up might be beaten—sometimes so severely that they died of their injuries. However, elderly women were much less likely to receive this kind of treatment, not so much out of deference to their age or infirmity as out of fear of the public uproar it would create. Some were so vociferous and eloquent that their husbands would not even argue with them, let alone beat them. These women reprimanded their spouses at the top of their lungs before a rapt audience of fellow villagers, young and old, who took delight in hearing an imaginative catalog of the unfortunate husband's sexual inadequacies.

From India, William Rowe (1961) has provided reports of de jure patriarchy and de facto matriarchy in which the older woman, sometimes known as the *malikin* (the queen) asserts real authority despite her lack of formal entitlement. Thus William Harlan (1964) discovers that, despite his ascribed status, the older man's prestige in the rural Indian family is in fact upheld by his wife, who works behind the scenes to insist that his edicts be obeyed. In this situation, the wife acts as interlocutor between the male head and the rest of the family. Without her, the aging father is rather content to lapse into passivity. It is the wife who brings issues that concern her to the old man's attention, quietly pushing him to act on them.

Moreover, the older man's titular power is not only manipulated by older women, it is also granted by them. They are, in effect, the kingmakers, and this power extends to the matriarch's younger successors. Thus should the older wife die before her husband, her mediating role in the Indian family is taken up by the wife of the oldest son. However, unlike the deceased mother-in-law, the daughter-in-law does not act in her father-in-law's behalf; instead, she makes a de facto transfer of domestic power, becoming the influence broker for her husband. This transfer of male power from fathers to sons does not result from oedipal rebellion, from male acts of symbolic patricide; rather, it is mediated by the women. They decide which man—father or son—shall be the ostensible patriarch.

When the older Indian woman rules the family through the office of the man, either her husband or her son, she conforms to both external (social) and internal (developmental) imperatives. She obeys the culture's convention of patriarchy and the inner, developmental push toward matriarchy. But this compromise solution is not exclusive to Asia. In North Africa, the socially sanctioned patriarchal arrangements of Muslim village society are not sufficient to check the unofficial matriarchy of later life. In rural Egyptian villages described by Soheir Morsy (1978), the elderly male head of the household often turns over to the wife his power to regulate the division of domestic labor and the distribution of resources within the extended family compound.

Other reports, from other regions, belong in this category. Fatima Mernissi (1975) tells us that the Moroccan matron, because of her privileged role in giving and receiving information about women, has "tremendous power in deciding who is going to marry whom. It reduces the man's decision-making role significantly." From Burma, Rustom (1961) reports that older Burmese women, particularly widows, become the heads of their households and are responsible for the management of family affairs "as long as they remain interested."

Summarizing these trends, Simone de Beauvoir (1972) observes that sex differences phase out with age; therefore, in Balinese cases, older women can occupy hitherto exclusively patriarchal slots and govern entire households.

The Old Woman as Mother-in-Law

When de Beauvoir tells us that old Balinese women may rule entire households, she is alluding to the matriarchal power that comes to the old woman as she acquires, through her son's marriages, a corps of daughters-in-law. The following vignettes give a sense of the relatively standard ways in which this power is gained and exercised, over a wide range of regions and cultures.

We have already learned of the elderly Asmat woman and her crescent powers. These are directed not only against the aging husband but also against the daughters-in-law brought into the household by her married sons. Van Arsdale's report of these arrangements read very much like other accounts from societies that, like the Asmat, maintain close-knit extended family structures. Van Arsdale writes:

> As their children began having children, the wives came to spend a larger proportion of their time inside the thatched family hut directing the labor of younger women, either daughters or junior wives (if the marriage was polygynous), giving them curt orders about cooking, childcare, fishing, and gathering. Sexual activity probably diminished as women neared "old age," but in polygynous marriages the senior wife compensated for her husband's neglect by reprimanding the junior wives for even the smallest inadequacies in their performance of household tasks.[2]

Residing on India's northern rim, the younger women of the Yusufzai Pakhtun are considered chattels by their fierce, patriarchal husbands, themselves the leading warriors of the wild Afghan border land. Yet Charles Lindholm (1981) found that women bear their subjection with patience:

> The wife's solace is hearing the tribulation of other women and anticipating the power she will wield in her later years. By that time, her husband will be a tired old man, without the energy for fighting, her sons will be grown, and their wives will be living in her house under her rule. She will control her domestic sphere like a real matriarch, and the *purdah* compound, her former prison, will become her court.

As we saw earlier, Hindu men, particularly those of higher caste, are enjoined by their religion to live as responsible householders during the period of active parenting, but to give up material strivings and ambitions in favor of meditation during the subsequent "forest" or "holy hermit" stage of life. But while the *vedas*, the religious codes, enjoin women to remain forever under the dominion of their husbands, in actuality the older woman can inherit the social influence that her husband has given up. Paul Hiebert (1981) writes:

> The father may have the leading role in society and in religion, but it is the mother who has the closest relationship with her sons and daughters.

In the end it is the mother who is revered most in the home and remembered most lovingly.

A woman reaches her best years when she herself becomes a mother-in-law and manages an extended household. Now she is cared for by her daughters-in-law and grandchildren. Now it is she who delegates housework to the daughter-in-law and goes out to shop or to sell vegetables or other products in the market. Under less pressure to spend as much time as possible pleasing her husband, an old woman has more leisure than before to visit in the courtyard with other old women, or to play with her grandchildren while their mother works. And even when the mother is free, Grandmother has first right to the children's attention.

Older women are also expected, under Hindu doctrine, to withdraw from mundane duties, to reflect on the next life, and to transfer household duties and powers to their daughters-in-law, as these come into the home. However,

In practice an old woman will generally delay giving authority to her daughter-in-law as long as she feels physically and mentally capable of handling affairs. She is glad to let the younger woman do the menial tasks but reluctant to relinquish the management of the house. The result is frequently a growing tension between the two that may not end when the older woman finally has to give up control because of disability or illness.

A full description of the Indian woman's metamorphosis from daughter-in-law to mother-in-law is provided by Manisha Roy (1975), whose report is based on fifty detailed life histories taken from upper-class Bengali women. The disparaged young wife, closeted by her life in purdah among strangers, eventually becomes

the *ginni-ma*, the matron-mother, a status that indicates the climax of a woman's life. Now she is important in her own standing. . . . She has served and obeyed the elders; it is now her turn to be respected by all, including the men in the house. . . . She is no longer a woman who can be looked at with desire or pity, but she must be respected. The keys that open and lock several rooms including the kitchen, the pantry, and the storeroom symbolize her command and authority in the household— something her mother-in-law and older sister-in-law once enjoyed.

Mernissi also finds that this special mother-in-law power is handed down, across generations, to those under her control and tutelage. Thus the teen-age Moroccan bride defers in ritual fashion to her mother-in-law and in exchange is instructed and protected by her. When in her turn she becomes a mother-in-law, the sons' wives defer to her; the marriage of her sons leads not to separation from them but to an increased claim on their love. The mother-in-law's household power is symbolized by her exclusive possession of the key to the storage room where food is kept for the entire extended family.

The Matriarch–Oldest Son Alliance

We have seen the older woman, surrounded by her retinue of young daughters-in-law, expand into the power vacuum left by the aging husband. But there appears to be yet another element in the matriarch's upward mobility, one that generally goes unreported. This elusive factor, glimmering through all the reports, has to do with the role of the adult son, usually the oldest male child, in catalyzing the familial role changes documented here.

The conventional wisdom in gerontology holds that the strongest bond of later life is between the aging mother and her daughters, excluding men. But this impression is based largely on parochial data from U.S. subjects. The older wife in traditional societies, however, has a male ally: She moves to power in concert with the oldest son, with each party taking over allotted portions of the old father's social powers. Through marriage, the son attains adult status within the family, while his mother acquires a potential rival—and a potential servant—in the person of the daughter-in-law. If the mother can retain some emotional hold over the son even as his affections shift toward a wife, then she and the son may share dominion over the family, with the mother becoming a senior adviser, an éminence grise who works her will in the family via the son. The daughter-in-law then becomes something of a vassal to the mother-in-law, thereby enhancing the senior woman's scope and powers. In this scenario, the son's marriage is fundamental to his own as well as the aging mother's advancement within the family.

This recurrent theme is particularly vivid in the African reports. Thus the marriage of her son fosters the elevation of the mother-in-law among the African Bemba. As Richards (1956) reports, the "shy and submissive girl eventually becomes the imperious and managing" older Bemba woman, who orders the activities of her younger kinsmen and kinswomen, as well as the distribution of food, from the shade of her veranda. Indeed, the power of the older woman could even reach beyond the household precincts. Senior Bemba princesses ruled whole districts of their own.

Spencer (1965) reports that older Samburu women acquire domestic power through alliances with their sons, the warlike, rebellious

Moran, against the depredations of their older husbands, who have by this time usually found a younger wife. In part, the mother's hold over her sons is based on her power to level potent curses against younger men. Thus the resentment and rivalry of the older woman is not confined to her sons' wives or to her spouse but is also aimed at younger men, even including her son.

Indeed, throughout Africa older women are bonded to younger men as allies, antagonists, or both. Thus Nadel (1952) tells us that the Nupe of Nigeria recognize two kinds of supernatural power: good power, used by men to combat witchcraft, and deadly power, used by the older women (who also engage in trade) to attack younger men. This particular variation suggests that older women may enact their rebirth as a "man" through alliances with virile younger males or through direct competition with them.

Recent findings by Robert LeVine (1976) among the Gusii of Nigeria suggest that this mix of feelings includes both erotic and rivalrous features. Postmenopausal Gusii women refuse to continue sexual relations with their husbands once they are past childbearing age; women over 45, if they have married sons, often reorganize their lives around their coresiding sons' wives and children, and subsequently have little to do with their husbands. In effect, they join their son's household as a kind of senior wife, even as they abandon their husband's bed.

In response to this defection, older Gusii men sometimes elect for polygyny, taking a younger wife in order to escape the sharp tongue of the postmenopausal senior wife. LeVine adds:

> There is also the continuing belief that a monogamous wife is likely to become proud and disobedient to her husband, a humiliating state of affairs. One man about 50 is said to have taken a second wife in response to taunting by his age mates that he would not dare to so defy his strong-willed first wife.

Frances Cox (1977) finds a like symmetry in male and female role change among the Ngecha, a herding tribe of Kenya. Like LeVine, she finds that older men develop "maternal" sentiments toward their herds, and she notes that one informant kept several sheep in his house, as a replacement for children. But as for women, "while their male counterparts are known for their quiet wisdom and humility, the women elders are sometimes accused of flaunting their independence, of being overbearing and proud." Although her ostensible control is confined to the younger females of her husband's clan, the older woman might attempt to move a young man through his wife. Again we see the commingling of love and rivalry in the attitude of the old wife toward her grown son. Cox notes that the older woman, jealous of her daughters-in-law, will test and flaunt the bonds between herself and her son by joking with him in the young wife's presence. But her rivalry is two-pronged; it is toward men as well as women: "However, in daily relations with her son, she is likely to have the reputation of behaving in a 'hard and all-knowing' manner, of being critical and offering unsolicited advice."

Family life in Sudanese Africa is similarly organized around the aging mother-in-law. Hayes (1975) observes:

> The older women achieve a status more closely resembling that of men. They have influence and authority over the daughters-in-law of the compound, as well as their own daughters still living at home. Mothers are greatly respected by their sons, and sons often have closer emotional ties to their mothers than to their patriarchal fathers. Grandmothers are respected as fathers.

Indigenous African men as well as foreign anthropologists have noted the special bonding between grown sons and older mothers. Here, as quoted by Cox, an old Ngecha man complains of his special isolation:

> I suffer from loneliness, a sense of lost manhood. My life is shrinking with age. Women do not suffer this loneliness, for sons feel much more sympathy toward mothers than toward fathers. I wish I could be like my grandfather who had eight wives, or my father who had four. But I have only one and I am lonely.

Colson and Scudder (1981) have also found among the Tonga signs of the special relationship between the aging woman and her grown son that we have already noted in other parts of Africa. Tonga widows are "inherited" by the dead husband's brother; if the wife is still fertile, a genuine marriage may develop. However, the postmenopausal wife, who cannot give him children, has only a second-rate, chattel status in her new home. Under such circumstances, the authors say,

> Aging women frequently leave inherited husbands to live with sons; frequently they leave them to live with other kinsmen. They do not expect much tolerance from daughters' husbands and say it would be foolish to live with a daughter despite the matrilineal rules of Tonga society. Elderly women feel greater security in their relationship with

sons. Even if their original husbands are still alive, they may choose to separate from them to move into a son's homestead.

The basic sexual architecture of the three-generational extended family that declares itself so bluntly through the African data is discovered again, essentially unaltered, in Asia. Turning first to China, Yap (1962) reports that when the aging Chinese husband retires from the management of the family, he concedes the control of its external affairs to his oldest son, and of its internal, domestic affairs to his wife. She will, for example, decide all matters relating to the marriage of her daughters.

As Margerie Wolf (1974) and Ch'ing K'un Yang (1959) independently observe, there is, in the Chinese family, an implicit but nonetheless powerful alliance between the grown son and his aging mother. This father-mother-son triangle announces itself most clearly in Taiwan, as seen in the account of Stevan Harrell (1981):

> As the Taiwanese woman grows older, her lot improves. As a young daughter-in-law, recently married into her husband's family, she is the lowest of the low. Her mother-in-law resents her claims on her son's affection and assigns her the dirty household tasks that she herself had been assigned by her own mother-in-law a generation earlier. Even if her husband genuinely likes her, he should not pay her too much attention.
>
> However, as she matures, she undergoes several changes in her social position that directly affect her status and comforts as an old woman. First, she becomes mother of a son, either by birth or adoption. This is significant both because she gains a modicum of respect from her husband's family, and because she now has a lever, *a potentially powerful adult male whom she will eventually attempt to control* [italics added]. . . . When her son marries, she herself gains a daughter-in-law to whom she can in turn assign the heavier work around the house and at the laundry ditch.

Lloyd and Susanne Rudolph (1978), in their study of a typical upper-class Rajput Brahmin family, give us a more detailed picture of an older woman gaining ascendancy through the marriage of her son:

> It is the presence of Amar Singh's bride in Jaipur that gives his mother great influence and authority over him. For 13 years she had no direct hold; now he moves into her ambit, and that of other family members and the servants. Their expectations and sanctions cannot be ignored; they shape his behavior as husband, son, father, nephew, and master-to-be. His mother, by comparison with mothers—and mothers-in-law—in the three related households, offers relatively few difficulties. They clash over what she regards as his excessive attendance over his wife. *The*

171

conflict is more over power than love—who is to have the predominant influence over the master-to-be of the household [italics added].

Here, close "clinical" observation has again highlighted an important motif in the mother-in-law/daughter-in-law struggle: The mother makes her bid for household power through the son, whose marriage has elevated him to manhood and placed him at the center of the household. In order to secure her supremacy, the older woman must downgrade the competing influence of the son's wife, her daughter-in-law. Robert Levy (1977) provides another set of confirmatory Indian observations, from the Newars of Nepal.

These massed reports from a variety of distinct and independent regional cultures—Africa, China, and India—converge toward a strikingly similar picture of role reversals in the three-generational extended family. Typically, these follow two pivotal, developmental events: the eldest son's attainment of sexual and social maturity through marriage, and the older woman's attainment of a more "masculine" virility at the end of her procreative period. Both advances are in phase with and perhaps require a third nuclear event: the voluntary abdication of patriarchal power by the aging father. It is as though the oedipal strivings that exist only in fantasy in early years finally (except for their specifically murderous and sexual goals) reach fruition in the adulthood of the son and in the postadulthood of the aging mother. In keeping with the terms of oedipal fantasy, the father abdicates (though without having to be forced) and the son inherits all the available women. Splitting his affections, he reserves sexual lust for his wife and more reverent feelings for his mother, as whose consort he rules the family.

The instances here cited, covering as they do a broad spectrum of cultural and physical types, suggest that we should rethink some basic assumptions in developmental psychology and psychoanalysis. For example, some revision of the standard psychoanalytic view of the Oedipus complex and its fate seems called for. Generally, it is assumed that the complex of oedipal motives are a feature of early life, dangerous fantasies that, in the course of normal development, come under repression and emerge only in disguised and sublimated forms in later life (as in the search for a mate other than the father or mother). In this version, it is only neurotics who carry the fantasy in its unmodified form into adulthood. However, it appears that this classic formulation is based on a special case—the fate of the Oedipus complex in the nuclear family that characterizes the modern city, rather than in the extended family that characterizes the premodern village.

In the setting of the extended family, the Oedipus complex is more than a personal myth; the oedipal fantasies are not an impossible dream but a rehearsal for adult and postparental arrangements that make very good sense in the small community and that mobilize the developmental potentials of all parties concerned. Thus the father does not fight the ascendant son but preserves his own and the general social tranquillity by moving out of his command post and toward a special relationship with the gods, meanwhile leaving the field of practical family administration clear for his newly matured heir and his newly "masculinized" wife. In this form, the oedipal fantasy can be lived out, with beneficial personal and social consequences, because the original dream has been denuded of its primitively aggressive and erotic components. The son does not have to kill the father in order to achieve his majority within the family, and he does not have to have sex with the mother. The mother-son "marriage" is "pure," unpolluted by lust or by blood guilt. Any residual misgivings over these transfers of the father's powers may be handled through externalization, in which each party to the oedipal compact uses the other as a kind of "projective ecology." The son can hold the newly aggressive mother responsible for the father's "castration," and the mother can claim that she is advancing not her own aspirations but those of a favored son.

As in her earlier years, the mother is still identified with the aggressive powers of a vigorous male, only now it is the son of her own loins; in contrast to her earlier years, she does not identify passively with the virility of her husband but instead works actively, in public or behind the scenes, to increase the powers of her son, her own "masculine" extension. In this regard, it may be that the older woman within the extended family realizes two fantasies in later life—the dream of union with the oedipal lover, and the dream of being made whole, both feminine and masculine. The high morale of older women may in part derive from these double fulfillments.

The Older Woman as Witch

As we have seen, the surgent energies released by later-life female development have generally benign personal and social consequences, particularly within their proper ecology, the extended family of the traditional village. But Therese Benedek (1952) observes that, although fairy tales often conjure up the "kind, discerning, loving, and under-standing grandmother," even more extensive is the folklore about the vicious "bad old woman." This disparity in description no doubt has to do with the new, surgent thanatos that comes to share equal place with eros in the older woman's psychic constitution. Thus it may be that, as older women reduce their motherly tenderness and accentuate their aggressive side, they become particularly threatening and are consequently at risk for acquiring a new social "name": witch, sor-ceress—identities that are clearly based on the older woman's revived appetite, even zest, for aggressive behavior. Thus Edward Wester-marck (1926) reports a Moroccan folk explanation of the older woman's diabolic power, one that clearly refers to the later-life redistribution of aggression between the sexes:

> When a boy is born a hundred evil *jinn* (devils) are born with him, and when a girl is born there are born with her a hundred angels; but every year a *jinn* passes from the man to the woman, so that when the man is 100 years old he is surrounded by one hundred angels and when the woman is 100 years old she is surrounded by one hundred devils.

In support of this folk wisdom, the ethnography of witchcraft does indeed show us a clear connection between the older woman's thanatos and her reputation for sorcery. When her anger toward family members becomes manifest, she is then at risk of being called a witch.

Some accounts suggest that a reputation for sorcery might supply older women, in male-dominated societies, with alternate routes to status and power. For example, Van Arsdale (1981) reports that the elderly women of the New Guinea Asmat could not usually attain formal status to equal the powerful old *tesmaypits*. However,

> Those who had established themselves as curer-sorcerers were both re-spected and feared. They were called *namer-o* because they derived their power from a non-ancestral spirit known by that term. Men could be *namer-o* as well, but apparently elderly women *namer-o* were considered to be both more powerful and more dangerous.

Along these lines, there does appear to be some association be-tween the acquisition by women of status in a man's world, and a reputation for sorcery. According to Nadel (1952), Nupe (Nigerian) witches are almost always women, and "usually an older and domi-neering female, who would attack a young man." The witches are presumably led by the head of the Nupe women's trade organization.

Nadel points out that the postparental Nupe women can earn more as itinerant traders than their farmer husbands, thereby maintaining financial control over the family. In this case, the witchcraft accusation may rationalize and focus the men's sense of threat from their socially and physically mobile, financially dominant wives.

The same compounding, in Africa, of supernatural and secular power was noted by John Middleton (1953). Among the Kikuyu, he finds, in every district, a *kiama* (council) of old women who seem to have disciplinary powers over younger women and who are feared on account of their witchcraft powers. Just as older Samburu men are suspected of using sorcery against their young rivals, the *Moran*, older Kikuyu women are suspected of enforcing their control over young women via witchcraft means. So it is in Manus, where, as Mead (1966) reported, younger women have come to believe that any old woman, including a close relation, may, if provoked, curse them and make them sterile.[3]

Clearly then, the older woman is granted, by social convention, power over the daughters-in-law; by this token, she gains honor. But the matriarch can also bring irrational oedipal rivalries to this arrangement, and such a surplus of self-indulgence can earn her the reputation for witchcraft.

For example, in discussing the polygynous extended family of Africa, LeVine (1963) points out that tensions are inevitable between a daughter-in-law and the mother-in-law to whom she is subordinated. If the bride does not get along well with her husband's family, she may charge that her mother-in-law mistreats her, even that she practices witchcraft. While some young women manipulate this accusation in order to excite sympathy, LeVine nevertheless claims that "many young women are genuinely afraid of their mothers-in-law, and suspect them of evil doings."

In sum, the African materials give us strong evidence of an oedipal triangle, with the son and mother again in league not against the father but against the daughter-in-law. Apparently, there are two oedipal victors in the postpaternal phase of family life: The son supplants the aging father, and the mother wins out over her particular rivals, the daughters-in-law. But while the fathers and sons can enjoy a post-oedipal comradeship, the mother-in-law acquires a reputation in the black arts. The reasons for this difference are now clearer: The old man has voluntarily abandoned power, to become the "angel"; but the older women, having grappled for power, becomes the "devil."

175

Thus the matriarch can become the equivalent of the "bad old man" of the Comanche, who uses sorcery as an expression of his unrelinquished rivalry with the young.

So it is not her aggression alone that puts the old woman at risk; it is the specifically masculine toning of her aggression that makes her eerie. When the matron becomes "phallic," she may be venerated as "manly-hearted," but she can also be desecrated, as a witch.[4]

Some observations by Geza Róheim (1930), based on his field work among the Australian Aranda, bear this out. Noting that the Aranda words for old woman, *arakutya knaripata*, mean literally "woman father," he adds, "the old woman partakes of the qualities of maleness; she is like the witches of European folk-lore, a sort of condensation of father and mother."

Westermarck (1926) offers this generalization: "Old age itself inspires a feeling of mysterious awe, which tends to make the man a saint and the woman a witch," and documents it with his notes from Tangiers, a city where the saying is, "An old woman is worse than the devil; nay, the devil himself is much afraid of her: She bottles him up." Even the "father of lies" fears the "woman father."

The Older Woman as Evil Spirit

This special fear of older women does not end with their death. For some African peoples, it even determines their spirit status. Max Gluckman (1955) writes of old Zulu women that "as ancestral spirits they were capriciously evil, while men's spirits brought only merited misfortune."

Similarly, among the African Tallensi, the fear of the aging mother extends to her ghost. Meyer Fortes (1949) notes that the spirits of dead parents are more persecutory than protective and that this "distortion is most striking in the case of the mother's spirit. More than others, the spirits of female ancestors are believed to be especially hard, cruel, and capricious."[5]

In Asia, as well as Africa, the fear of the older woman's witchcraft can extend to the afterlife, to her spirit manifestation. John Batchelor (1928) finds that "the [Japanese] Ainu fears nothing so much as the spirit, soul, or ghost of a dead female ancestor." He found that even before their death, old women were particularly feared by children, and that the ghost of a deceased old woman had a particular power to do evil.

Indeed, the fear of the old woman's ghost was so great that the

Ainu would burn down the hut of the oldest woman in the family when she died, for fear that her spirit would return to the house and, out of envious malice, bewitch her offspring and extended family.

The African accounts suggested to us a direct association between the heightened status of the old woman and the fear of her witchcraft potential; the same linkage is found in Papua New Guinea, as reported by Leopold Pospisal (1958). He discovered that senescence has opposite effects for men and women among the Kapauka. It leads to an eclipse of the man's independent position by making him a member of his son's household and by decreasing his social status, but it has an opposite effect on the status curve of the woman. The older she grows, the more she becomes emancipated from the powers of her husband. Because he respects her age, she is no longer beaten or scolded. This respectful stance is generalized, extending to the younger generation, who try to please her in order that, after death, her spirit will be kind and helpful. Men lose their magical power, as shamans or sorcerers, to do harm after death; but the danger from *meenoo*, the female ghoul, not only persists, but grows greater as the woman's body approaches the decease that will free the *meenoo* to settle in another woman. The elevation in the woman's status curve begins with the onset of menopause, when the eating taboos against women are lifted and her rights in this respect equal those of men.

Death may take away the potentially malign powers of the man, but it releases those that have been latent in the woman, and, as noted, the old woman can be viewed ambivalently, as a holder of "bad" as well as "good" powers. However, in some settings, as reported above, the full appreciation of her bad powers may come about only after her death.

Primitive and, particularly, oral sadistic features become even more evident as we move away from reports of real women, in real communities, to consider the representations of the old witch in folklore and mythology. In these accounts (as in the Hansel and Gretel story), the old woman is an eater of human flesh. For example, this folk tale cited by Verrier Elwin (1958), from the Sherdukpen tribe of the Indian northeast frontier, describes the land of Sirinmus, beyond the Himalayas, a place inhabited only by women:

> If a man went by accident to that country he was lucky if the young women caught him, for they would keep him for only a week and if he survived their passion they would send him home laden with rich gifts.
> . . . But if such a man fell into the hands of the old women, they would shut him up in a house of stone and close it with a stone door. They fed

him with the best of food until he grew fat. When he was ready they filled a great pot with water and put a fire under it and when the water was boiling they threw him in. When he was properly cooked the old women ate him with loud cries of pleasure.

Another folk tale, collected by Elsdon Best (1925) among the Maori, concerns the Rururhi-Kerepo (blind old woman) and her misbehavior with five girls whom she finds in the forest. Having first demanded that they call her "aunt," Rururhi-Kerepo made all the girls climb a tree:

> When they had done so, she called out, "Oh, my nieces, how nice you look up there; I could eat each of you at a mouthful." Then she shook the tree violently, crying out, "Drop off! Drop off!" As each girl fell from the tree, the old woman seized her, bit her head off, and ate her body. Later the men who were with the girls came searching for them and found only their heads. They discovered the ogress and, after a fight in which one of the men was eaten, slew her.

Since the mythic narratives are not constrained by reality, they give us a better sense of the primitive emotions that are projected onto and lived out vicariously through the old witch figure. Earlier we saw that older women can become counterplayers, with their sons, in a late-life oedipal drama, but in these myths we see that older women can also become the screen for even more infantile, primitive projections. In them, young women are depicted as essentially benign. The Sirinmus maidens may demand sex from a man, but they will reward his performance with gifts; it is only the old women who are cannibals. By the same token, Rururhi-Kerepo demands that the young Maori women greet her as an aunt, as a motherly figure, before she eats them. Though she is first identified as a maternal person, in the course of the story the old woman becomes a devourer, rather than a preserver of children. In effect, Rururhi-Kerepo represents a generic danger associated with mothers: the danger of merger, of losing identity in her being. The primitive wish to merge with the mother, to take her in orally, is transmuted, reversed, to become a frightening aspect of the mother: "*She* wants to eat *me*." Thus the mother is always seen ambivalently, reflecting threats as well as promises; the discordant aspects of this ambivalence are split from each other and portioned out among the representations of young and old women. Young women attract the positive aspects of the ambivalence, to become figures of loving nurture, while old women—perhaps because of their more obvious aggression—attract the negative aspect, becoming the cannibalistic witch who will force, with her teeth, the feared but unconsciously wished-for merger.

The Older Woman as Role Creator

Thus far, we have looked at the old woman of folk-traditional societies and traced out the various forms that she achieves—as matriarch, priestess, or witch—within a relatively universal, recurrent social form: the matrilocal or patrilocal extended family. While the predicted growth of the old woman's powers does take place in these settings, cultural anthropologists might argue that this outcome does not reflect development but is instead the consequence of the social nurture or "opportunity structures" provided by a special social format—the traditional folk society. They might argue that it is only in such settings, and because of them, that the older woman's metamorphosis takes place, through events and traditions that are explicit, culturally recognized, and culturally sponsored: The office seeks the candidate.

Nevertheless, we do find instances, even within the traditional folk society, in which the older woman's personal or social maturation is not socially sponsored. Such cases allow us to test the proposition that older women will proactively, on their own hook, assemble networks that will sponsor their maturation, even when these are not ready-made for them.

A good example is provided by Michael Lambeck (1985), who studied the Mayotte community, on an island off the Swahili coast of Africa. By contrast to many of the folk-traditional settings that we have reviewed, Mayotte society is not characterized by strong extended families. Instead, families are amorphous, shifting in their boundaries and often held together by impermanent affections rather than by socially enforced norms of mutual responsibility. Divorces are easy to obtain, and some women have had as many as twenty "marital" liaisons in the course of their lives. Just as parent-parent ties are not fixed, so it is with parent-child ties; women easily "borrow" each other's children through more or less permanent adoptions, thereby building up "flocks" of young to provide companionship and domestic service. In effect, women, particularly old women, use this fluid system not to escape parenthood but only to make that state more rewarding. Lambeck writes:

Because child transfers are easily arranged, no one need remain with or without children should they desire otherwise. Childrearing is an activity which women can pursue throughout their lives; alternatively, they can avoid it. In the circulation of children, middle-aged women play a major role, although men, women, and the children themselves may have an interest in the matter. . . . "Fosterage" helps to cement marriage alliances and maintain its links despite divorce. Women can also restructure kin ties through the establishment of "fictive" kin and reinforce these links by means of child transfers. The more children that she raises the more can a woman participate in arranging marriages and subsequent child transfers. As time passes she becomes the center of an ever expanding *mraba* ("family") with ever greater links with other members of the community. By maintaining an active interest in children and grand-children she can in turn expect their respect and support.

In this telling example, we can see the older woman actively con-structing the extended family networks that, in their turn, sponsor her own development as a powerful personage, both within the family and outside of it. In Mayotte, the extended family is not ready-made for the older woman as an "opportunity structure," demanding her services in later life. The office does not seek the candidate; instead, the candidate creates the office by actively generating an extended constituency that is centered around her and takes its direction from her. Again, we see the surgent aspect of the developmental series at work, as it drives the older woman toward the next step in her mat-uration, that of discovering and even creating the "facilitative" ecol-ogy of an extended family.

The Older Primate Female

Finally, for those who would argue that ascendant older women are merely responding to social prompting and instigation, I cite some recent studies of aging subhuman primates. These indicate that the virilization of the older primate female may be general not only across human societies, but across primate species as well.[6] Susan Blaffer Hrdy (1981) conducted field studies of the langur monkeys and of the very successful macaques (successful in the evolutionary sense: they are distributed across a wide range of varied habitats), finding that, like the majority of younger human mothers, the maternal ape is de-voted almost exclusively to the care of the latest infant, the one still clinging to her fur. She is at the same time relatively inoffensive in her dealings with other adults. Thus, during the period of intense parenting, she allies herself with dominant males and trades sexual access for their physical protection. However, in the postparental years, a striking change occurs that parallels our cross-cultural observations among human females. Hrdy observes that the ways in which older monkeys support younger animals "seem to vary with the sex of the animal and the situation of the group. Males generally bow out, leaving older females to intervene actively in the fate of their descendants."

In effect, female primates play two distinct parts in regard to procreation. They provide physical as well as emotional security, though—consistent with the exclusivity of these roles—they play them out sequentially rather than concurrently. As Hrdy reports, "When a troop of Langurs is threatened by dogs or humans, or by encroachments upon its territory by other Langurs, it is typically the adult male *or the oldest females* [italics added] who leave the rest of the troop to charge and slap at the offenders."

Similarly courageous and persistent defense of younger relatives by older females has been documented for Japanese and Rhesus macaques by Jennifer Partch (1978). According to her, when troops are in conflict, postreproductive females are more likely than breeding mothers to chase or even attack intruders. Partch has recorded 572 separate instances of protection or defense of infants by postreproductive females, on occasions when breeding females did not take up the defense of their own offspring.

Clearly then, postreproductive female primates adopt the defender's role that we had thought to be reserved for young and vigorous males. The breeding mothers are almost exclusively providers of emotional security; but, like males, the postreproductive females move at times of danger not to the protected center of the troop but to its outer defense line. Even in the absence of social pressures as we know them, we find the older nonhuman female, who lives in a non-cultured state, showing the same role shift, the same expansion into hitherto exclusively masculine positions that we found among elder human females, who live in a cultured state. Culture may play a part in determining the particular, parochial forms of role involvement, but it cannot account for the older female's striving, general across human societies and primate species, toward masculine roles and statuses.

The later-life emergence of the older female is ubiquitous, across species as well as cultures. Accordingly, it should not surprise us that Frances Purifoy, Lambert Koopmans, and Ronald Tatum (1980), who investigated the fate of androgens and estrogens in later life, find that the "male" hormones, serum testosterone and free testosterone, increase their concentrations in the blood samples of aging women to a highly significant degree. Meanwhile, free testosterone, the "prototypical" male hormone, falls off for males in later life, a finding again consistent with our predictions. To round out the picture, these authors report a finding of "relatively stable or increased estrogens in older males." In sum, there is a rise of "masculinizing" hormones in older women and a concomitant rise in the "feminizing" hormones in older men, and this phenomenon, though studied in only one society, is consistent with the cross-cultural reports and probably accounts at least for the surgent, biopsychological phase of elder matriarchy across all human societies (and some primate species).

Surgency, Facilitation, and Sculpting in the Older Woman's Development

Reviewing the cited ethnographic data from traditional folk societies, some regularities emerge that again point out the broad range of roles and settings in which the older woman's later-life metamorphosis can take place. We see them becoming more bawdy but at the same time (and sometimes in the same places) more powerful in the religious rituals of their people.

We also see older women coming forth as matriarchs of the extended family, and crafting new relationships with their grown sons, relationships in which each partner becomes the active principle in the other's ascent.

At the same time that she is gaining secular power in ritual, family affairs, and community politics, the older woman may also be gaining supernatural power, either as a witch or, in postmortem social existence, as an evil spirit.

I have already commented on the broad range of the older woman's later-life investments, but this brief review makes it clear that the family, in particular the extended family, is the major arena, the psychosocial ecology that is most directly reciprocal and facilitative to the surgent potentials carried by the older woman herself. Thus 47 percent of our references point to the older woman's ascendancy within some version of the extended family, and these references come from all major areas of human habitation. If we add in those citations that deal with the old woman's vengeful sorcery toward other family members, more than 50 percent of them refer directly to the empowerment of the older woman in the family setting.

While the older woman can find or create a developmental anchor in extrafamilial settings, across the planet the extended family does seem to provide the natural ecology that is best suited to the developmental task of later years: It brings about the transformation of the matron's eruptive potentials into formed, stabilized executive capacities.

I have cited data bearing on the fantasies, appetites, and psychological capacities of older women, on the ways in which they deal with their intimates, and on the kinds of social roles that they perform in societies that invite their contribution. I have extended the comparative base beyond cultures, to include nonhuman primates. In effect, we have reviewed data from the various levels called for by a developmental analysis: The psychological data, particularly those that tap unconscious fantasies, give evidence of the phenotypic/eruptive phase; the data on the older woman's intimate relationships give evidence of the proactive/facilitative phase; and data concerning the "official" roles of older women give evidence concerning the reactive/ sculpted phase. From the psychological data, we learn that older women have raw potentials, appetites, and energies that can expand into many channels, to fuel many behaviors across a wide spectrum of executive roles. From the data that bear on close relationships, we see the older woman pushing aside impediments—whether these be her husband or her daughters-in-law—and finding congenial allies (as in her son) who advance her purposes in the family even as she advances theirs. Finally, the reports on her formalized social roles testify to the sculpted/reactive phase of development. In most settings, the older woman is sculpted toward being the matriarch of the extended family; in others, she becomes the priestess or politician, and even the dreaded witch, the negative version of the priestess. But whether aging women become leaders, witches, or both, a common factor underwrites these sundry transformations, and that is the general increase in their powers.[7]

As we have seen, power always wears a double face. In most cases—as when they become matriarchs—older women demonstrate their endowment of good power; in other cases—as when they are reputed to be witches—older women show the bad face of power. But in the latter as in more innocent cases, their potencies do increase, giving evidence of a surgent developmental awakening. Thus the transcultural variety in the older woman's role acquisitions registers the shaping effects of imposed forces—for example, parochial norms and usages—but the generic, across-the-board increase in sheer personal vitality registers the constant, invariant presence of the geno-typic or eruptive phase of development.

The fact that the older woman's newly liberated energies have been transmuted and sculpted, via their impact with the social environment, does not negate their ultimate biological, developmental origins. A rainbow shines with many colors but is everywhere composed of the same water vapor; only as it refracts light does that moisture become visible as a many-hued spectrum. By the same token, the energies of the older woman break into the public light, to become continuously visible only as they intersect with the social domain and are refracted into various distinctive roles and usages. Finally and paradoxically, it is the wide variety of roles invigorated by the older woman that signals the presence of a unitary though protean energy, rather than many unrelated and fortuitous social phenomena.

The developmental hypothesis, that men and women move jointly toward a later-life condition of androgyny, has been reasonably established. In the next chapter, I will consider the evolutionary reasons, having to do with the requirements of human parenthood, for this profound reorganization of the human psyche in the later years.

8

AGING AND
THE PARENTAL IMPERATIVE

I ask you, my friend,
What ought a man want
But to sit with his wine
In the sun?

My neighbors all come to talk over the new,
And to settle the problems of state.
I've no taxes to pay on my house or my field.
I'm lucky, you say?
So I am!

My three sons?
Married, all of them;
Fine wives they've got,
And from the best families, too!
My daughters?
I've five of them, all wed.
Good husbands I found for them,
And every one rich!

So I sit in the sun
With my jug of old wine,
And I'd not change with the lords of the land!
—WANG CHI, ca. A.D. 600*

U p to this point, our text has been mainly descriptive, documenting the expressions of the three great relational modes: aggressive
(thanatos), affiliative/communalistic (eros), and narcissistic (omnipotential, omnisensual) as these vary by age, by sex, and by culture.

* Reprinted from *Poems of a Hundred Names* by Henry A. Hart, trans., with
the permission of the publishers, Stanford University Press. Copyright © 1938 by the
Regents of the University of California. Copyright © 1954 by the Board of Trustees
of the Leland Stanford Junior University.

Generally speaking, we have found that differences in age and gender predict to changes in these relational modes across a wide range of societies: As they age, men give eros priority over thanatos; aging women give priority to thanatos over eros; and both sexes, moving toward omnigratification, become more narcissistic. We have, in effect, described species regularities that point, quite decisively, to a developmental staging of the aging process.

But the work of establishing a developmental geropsychology remains uncompleted. At this point, I have demonstrated a developmental effect but have not yet specified its nature and causes. We are still faced with a crucial question: What is the dynamic of change? What engines power the developmental movement of individual lives along the tracks that we have mapped?

Species Freedom and Parental Servitude

Any species-embracing developmental sequence of the sort documented here must in some important respects be guided and orchestrated by genetic factors. These biological agents preserve, within the human gene pool itself, the evolved resources that characterize the hardiest members of our species. In considering some of the human properties that make for individual hardihood and species survival, I introduce an essential argument of this chapter: that individual and species success are both judged according to parental criteria. By these standards, the hardiest individuals are those who survive at least long enough to raise viable, potentially parental offspring; species success likewise depends on parental continuity, although across numberless generations. Ultimately a species can survive only when its parenting aptitudes, as genetically directed, ensure the survival and maturation of children who will in turn raise their own viable children to be successful parents.[1]

The human concentration on parenting is determined not only by our low offspring-to-parent ratio but also by the special vulnerability of human children. The centrality of parenting in human affairs (and in primate affairs, generally) is based on a paradox: It is an inescapable consequence of liberation, of the relative human freedom from the tyranny of "old learning," of inflexible, "wired-in" instincts.

By contrast with all nonprimate species, our children are born with very few instinctually guaranteed, reliable skills. They are blind, the nervous system is not morphologically mature, and save for the inborn capacity to orient toward and suck the mother's breast, the human neonate lacks a fixed, reliable behavioral repertoire. Instead, it has a large reserve of unformed cognitive and social potentials, each of which requires special tending, by special sponsors, if it is to develop from the original, diffuse state to its sculpted outcome as an executive capacity of the individual. In the human case, genetic determination chiefly accounts for the surgent phase of development, the arousal of diffuse potentials; special social nurture is required, at each subsequent stage, to complete the maturing process for any given capacity.

Human infants are tremendously vulnerable but also tremendously gifted; instead of fixed, inherited old learning to get them through infancy and childhood, we have awesome potentials for new learning stored in the neocortex. Our species has traded old, "pre-wired" brain for the *tabula rasa* of the neocortex, the result being a learning explosion that is registered in each successive generation and that has, in a few millennia, made our species the masters of this planet and its near-space environs. But if we are to reap the evolutionary advantage of the human cortex, the parent's matured functions must serve to keep the child alive until its own matured brain can come on stream to embellish the general store of new learning. Our species' freedom from the restrictions of old learning commits the young to a long childhood and the human adult to a long period of parental servitude. Mothers and fathers must meet the child's need for physical and emotional security until their offspring are ready to supply these requirements for themselves, and ultimately for their own children.

Parenthood and Adulthood: An Equation

The dialectical relationship between infantile freedom and parental constraint is registered by most successful human groups in that they bring about a clear identity between mature adulthood and parenthood. Indeed, contemporary American society is one of the few in which a major attempt has been made (outside of monastic orders) to split adulthood from the parental condition. Most human societies are like the Kota of India, in which David Mandelbaum (1957) found that a Kota man does not think of himself as fully human until he has had children, "until there is someone to call him father." Similarly, Levy (1977), who studied adulthood among the Newars and in Tahiti, found at both sites that "a responsible parenthood (the keeping and rearing of children) produces a shift in *informal social definition* to adulthood, confirming the earlier *ritual definition* at marriage" (emphasis in original). And Lowenthal, Thurnher, and Chiriboga (1975), who studied the psychological effect of important life-span transitions, found that even in our relatively contra-parental society,

> Family centeredness is a dominant theme in the protocols, and parenthood is the main transition envisaged by the young. Work, education, and marriage were viewed largely as a means to that end. Although we saw signs of expansiveness, experimentation, and interest in personal growth among the young men, they primarily hoped to emulate the styles of the parents' generation, and it would appear that they would eventually accede to the nesting inclinations which make up the major goals of the young woman.[2]

I became intuitively aware of the equation between adulthood and parenthood in the course of my field interviews. While the announced purpose was to study the psychology of later, postparental life, my male subjects vigorously impressed on me the importance that parenthood had played in shaping their adult life and character. At the study's outset, I did not interview my Navajo, Mayan, or Druze informants about parenting issues; nevertheless, in the course of my unstructured interviews, the parental motif asserted itself time and again. The interviews were open-ended, exploratory; they were not formed around my predetermined priorities but were shaped to elicit informants' subjective priorities and concerns. Thus unprompted, the parental theme in the lives of younger subjects (in their late thirties and forties) is stunningly clear: Time and again, younger male subjects linked their pleasures, complaints, and remedies to their situation as parents and to the welfare of their families. Predictably, in unstable as well as stable societies, men told me that they had been wild in their youth, but that marriage had shifted their character dramatically toward greater responsibility, selflessness, and moderation: "I used to hell around; I didn't care for myself or anybody else. [I see that you are not like that now. What happened?] You know how it is: I got married; I had kids."

Also striking is the degree to which younger men define both their pleasures and their pains in terms of their family's welfare. As we observed earlier, younger men equate contentment with good health and with a sufficiency of food—not for themselves, but for their dependents. Thus I ask a Highland Mayan father, "What is it that makes you happy?" His answer, fairly standard for his younger age group, is quick: "When there is corn and beans for my family. . . . When things are bad we always think, 'Where can I find work? Where can I find the pesos to buy corn and beans?' " And a Druze father in the same age range will tell me, again quite routinely, that he is contented when there is "peace in the home, my family is healthy, and there is peace in the village."

The main message from the middle-aged fathers was that marriage, and particularly parenthood, ended the fun and games of young manhood, and that these pleasures were largely unregretted. Many of us have heard and even rendered the same wry judgments on parenthood, but their repetition across the whole cultural range transforms a seeming banality into a human universal of some dignity. The transcultural ubiquity of such statements points again to the universal importance that parents, including fathers, attach to the fact of their own parenthood. In short, fatherhood seems to mobilize profound emotions and sentiments, and these find standard expression across disparate societies. If I make parenthood central to my own thinking about aging, it is because, without quite expecting to, I found it to be central in the lives of my aging informants.

I also found that parenthood, as a central human concern, sponsors standard comprehensions about parenting across cultures. While developmental psychologists may argue about the importance of early experience versus the "here and now," illiterate peasants did not think it strange that, in my life-cycle interviews, I focused on such matters as weaning, toilet training, sibling rivalry, or the impact of a stepmother on early character development. They already knew about the effect of these contingencies on a child's feelings, on the family's mood, and on the child's later development. I do not think, when it comes to parenting, that any reasonably successful child-rearing practice is inconceivably strange to any human group. Regardless of culture, my informants could discuss child-rearing arrangements other than their own (and including mine) with true psychological sophistication, agreeing as to the range of possible practices, from harsh physical discipline to laissez-faire permissiveness, and as to the developmental consequences of these various styles. As a powerful and standard experience, parenthood seems to enforce its own universal norms and understandings, despite the widest disparities in cultural belief systems. Like war or love, parenthood has its own stern, preemptive aspect; it is an experience that speaks to and reveals the common nature in all of us. And the compelling nature of the parental experience is demonstrated by the common understandings, concerning the ways of parenting, that this state kindles in those who have shared this experience, however much their lives may vary in other ways.

Gender Differences in Parenting

As a primary experience, parenthood also enforces not only common understandings but also a stern discipline. As noted earlier, parents foot the bill for our freedom from instinctual coercion, and for our opportunity, as humans, to create new intellectual forms, to create new social forms, and even to take control of our own evolution. Accordingly, while societies have different ideas as to the developmental end points that they want to achieve through their child-rearing methods, they nevertheless maintain common understandings about the basic needs that must be met by any child-care regime. And there is general agreement that parents will, as part of their general servitude, accept deep restrictions on their own needs and deep revisions of their own psychological makeup, in order to meet their children's essential needs. There also appears to be general agreement as to the nature of these basic needs: If it is to thrive by any developmental criteria, the vulnerable child must be assured of two kinds of nurturance: the provision of physical security and the provision of emotional security.[3]

The fact of gender captures and memorializes the processes of evolutionary selection whereby the necessary capacities were assorted by sex, so as to assure the provision, to children, of physical and emotional security. Let us first consider the masculine responsibility for physical security. In the species sense, there is always an oversupply of males, in that one man can inseminate many females, but women, on the average, can gestate only one child every two years during their relatively brief period of fruitfulness. The surplus of redundant males, those over the number required to maintain viable population levels, can be assigned to the dangerous, high-casualty "perimeter" tasks on which physical security and survival are based. The more expendable male sex, armed with large muscle and a greater store of intrinsic aggression, is generally assigned to hunt large game; to open, maintain, and defend distant tillage; to guard against human and nonhuman predators; or to raid other communities for their wealth.[4]

By the same token, the sex on whom the population level ultimately depends is less expendable. The sex that has breasts, softer skin, a milder nature, the sex that fashions the baby within its own flesh, is generally assigned to secure areas, there to supply the formative experiences that give rise to emotional security in children. Indeed, under conditions of guaranteed physical security, the mother's provision of emotional nurture to the baby can be seen as an extension of her intrauterine care to the fetus. The mother's extrauterine nurture to the infant consolidates its psychological structures, just as her own body once formed and consolidated its physical structures.

The basic division of parenting duties is coded by gender, and that code is understood by most human societies, dictating the assignment of men not only to warfare but to almost any "perimeter"-based activity. For example, George Murdoch's (1935) tables, based on ethnographic data from 224 subsistence-level societies, indicate that any productive or military activity requiring a protracted absence from the home—hunting, trapping, herding, fishing (particularly deep-sea fishing), the pursuit of large sea animals, offensive and defensive warfare—is performed almost exclusively by males. Activities carried out closer to home—dairy farming, erecting and dismantling shelters, harvesting, tending kitchen gardens and fowl—are sometimes exclusive to men, more often exclusive to women, but are in many instances carried out by both sexes. However, hearth-side activities, particularly those having to do with preserving and preparing food, are almost exclusively the province of women (and, as we have seen, the occasional old man).

Clearly then, it is not the capacity for hard labor that distinguishes the sexes but the site at which the labor is performed. Women can work, often harder than men, at labor that does not take them away from the domestic zones; but even idle young men, telling dirty jokes in some camp off in the bush, comprise a military force. These warriors on the periphery even in idleness guarantee the domestic center, so that women may carry out, in safety, their vital work. Just as the living cell requires a nucleus and a boundary wall, so does the small human society require its warm, affective core and its stubborn, flinty redoubts. If parenthood is to go forward, despite the usual human conditions of relative danger and scarcity of resources, then two distinct groups, usually assorted by sex, are required to maintain these distinct but coacting structures.

Cultural rules may carry some of the burden of maintaining these structures and of assigning men and women to maintain them. But the basic division of labor by gender is an aspect not only of human but of general primate nature. The lower primates share with the human species the protracted period of childhood vulnerability; this blunt fact of primate existence seems to have fostered an equivalent division, by sex, of parenting roles. The primate horde—particularly the more vulnerable ground-dwellers, such as the baboon—is protected by a defense force composed mainly of males who move out to the perimeter when predators approach and who form a scouting line when the troop is on the move. Females look mainly to adult males to provide physical security for themselves and for the offspring who cling to their fur, and the best fighters, the dominant males, have privileged sexual access to females. Their sexual primacy is not only enforced by the stronger males but is granted by the females. In effect, females bribe males, by the promise of sexual access, into using their superior strength for defensive and ultimately paternal purposes.

Despite the vast differences between humans and lower primates, there is a central commonality, having to do with the vulnerability and dependency of offspring, that brings about an investment in distinct parental roles, across species. In both humans and apes, adult life is circumscribed by the stringent requirements of primate fatherhood and motherhood. Among the larger primates, the demands of parenting dominate and define adulthood, as well as determining the allocation, by gender, of parenting roles. In early life, human development goes forward because offspring receive parenting; the thesis of this chapter is that the parental imperative orders not only the maturations that precede the reproductive years but also those that come about when that period has ended.

Gender, Early Socialization, and Parenting

If human development is organized by two great principles—being parented in childhood and being parental in adulthood—then the child's reproductive destiny should organize even its early experience and the forms of rearing that it gets. And we do indeed find that human socialization practices, as enforced by parents and teachers, underline surgent, genetic dispositions, preparing males for the wide-ranging perimeter roles of adulthood and preparing females for the domestic center. Thus Herbert Barry, Marvin Bacon, and Irvin Child (1957), having abstracted ethnographic data concerning early socialization from 110 distinct, marginal-subsistence communities, report a striking pancultural consensus: Despite the usual intercultural differences, societies recognize not only the gender distinctions of adult parenthood but also the gender distinctions of childhood. Standard parental requirements not only dictate men's and women's assignments in adulthood, they also give content and meaning to the play, the education, and the indoctrination of boys and girls long before their entry into actual parenthood. The gender roles of adulthood do not only entail particular behaviors; if these are to be predictable and trustworthy, they must rest on a sound base of psychological structures: tendencies and motives that are inculcated and quickened long before specific parental behaviors are required.

Barry, Bacon, and Child find that their sample societies routinely prepare males for the perimeter and females for the domestic center. For example, males are almost universally socialized toward achievement and self-reliance in a world they never made, the lands beyond the perimeter. The themes that underlie male socialization are most vividly dramatized in puberty rituals, in *rites de passage*. Whatever their specific form and choreography, these always involve some form of ordeal for the young, aspirant male, who—usually at the hands of senior males—is scarified, humiliated, frightened, or all of the above. In this fashion, the elder males test the courage and endurance of the young male. If the young candidate passes the test without crying, without in effect calling for his mother, then he has made it as a man. He has demonstrated that he already has or is likely to mature the qualities of endurance and courage that are called for on the perimeter. Whether as trader, hunter, soldier, rebel, itinerant merchant, or worker, the candidate for manhood moves from the inward, central location of the mother's world to the outward perimeter of his father's world. The puberty ritual marks the rebirth of the young male, from being his mother's child in the home to being his father's son on the perimeter. And in addition to strengthening his ties to the fathers, the young candidate confirms his ties to the "brothers," the age class of boys who endure the ritual with him. The puberty ritual thus strengthens male bonding and introduces the boy to the comrades, mentors, and father-gods who will be his companions and comforters on the road. The ritual gives him male allies, the special portion of the community that he can take with him, even as he leaves the larger community on his sorties beyond the perimeter, and away from the mothers.

Just as societies quicken the qualities that will make men perform reliably on various frontiers, so are women urged to cultivate the qualities that will fit them for their role as providers of emotional security, within the heart of the family and the community. The data compiled by Barry, Bacon, and Child make it clear that the central themes in female socialization across cultures are nurturance, responsibility, and, to a lesser degree, obedience. This is not to say that women are not potentially aggressive or that men are not potentially nurturing; but responding to the mingled social and species coercions, each sex amplifies only a limited sector of the potentials available to it. In their socialization practices, societies are guided by two main criteria. They fix on those psychological potentials that are native to the particular sex and that are at the same time germane to the parental roles that men and women will play out.

The Entry into Marriage

In effect, the socializing practices of stable societies undercut the individual's narcissistic illusions of omnipotentiality. By and large, they do not permit individuals to enjoy and amplify the full range of psychological possibilities, "masculine" and "feminine," that are potential within them. But men and women can regain externally, through mating, what they have lost internally, through repressive socialization. That is, men and women can rediscover outwardly, in their mates and sexual partners, those qualities that have internally been ruled off-limits.

Those self-qualities that are inconsistent with the gender identity change from bad to good as they cross the boundary of the self to be experienced as qualities of the other. Thus the wife or lover provides an external medium, a kind of projective screen through which men can recontact, appreciate, and vicariously live out the qualities of nurturance and tenderness that complete them, that restore their lost duality, but that can no longer be tolerated within the self. By the same token, women can discover and appreciate in men those qualities of aggression and dominance that are blocked from expression within themselves. They can admire and even sponsor male aggression (so long as it is directed away from themselves and their children), living out their own desires for exploit and competitive success mainly through their husbands' achievements.

The young woman gives up her more assertive strivings in the interest of domestic harmony and in the service of her future parental role, to be a provider of emotional security to children and of emotional comfort to the providing husband. If she kept title to these qualities, if she permitted free expression of her aggression toward demanding, draining offspring, she might end by abusing and damaging her children emotionally and even physically, as well as driving away the provident husband. Having relinquished the dominating qualities that could interfere with her parental role and be destructive to her children, she rediscovers them, and enjoys them, in her mate or lover, often choosing precisely those men who best depict the unruly potentials that she has surrendered for herself. This adaptive restriction—as managed by the representatives and institutions of culture—serves parenthood in two ways. Within each sex, it closes out psychological tendencies that could interfere with adequate parenting, and it leads individuals to seek their completion, their lost omnipotentiality, not within themselves but through procreant alliances with the heterosexual other. In order to regain the lost sense of omnipotentiality, men and women must seek completion through intimate liaisons, and these lead on, quite naturally, to reproduction and parenting.

The Chronic Emergency: The Onset of Parenthood and its Consequences

While the necessary sex-role training begins early in life, its consequences are not fully evident until the actual onset of parenthood. With the coming of the first child, there also comes a chronic sense of emergency, and a general mobilization, in men as well as in women, along the lines suggested above.

Prior to parenthood, and despite their early sex-role training, young men and women are allowed, in most societies, some freedom to indulge a wide range of psychological potentials. Thus prematernal women are often tomboys, flirtatious one day, actively competing with men the next, while young men, including prepaternal husbands, may live out the extremes of their nature toward violence on the one hand and tenderness on the other. Before parenthood, young men and women are allowed some freedom to live out their narcissistic strivings toward omnipotentiality—toward conserving, for the self, all possible potentials and options, no matter how mutually exclusive these might finally be. However, the entry into the condition that I call the "chronic emergency of parenthood" leads to an energizing, in young parents, of the structures that were laid down during socialization, as well as a muting of the claims toward omnisatisfaction in all domains. After children come, dedicated parents can never completely relax into self-absorption or self-indulgence. From then on, even rest becomes a nurse's sleep, the parent waiting for the child's cry or the alarm in the night.

Both men and women respond to the parental emergency by instrumentalizing themselves to meet parental requirements. Thus young fathers become an extension of their hardened, functional tools and weapons. In this service, they tame the extremes of their nature, deploying aggression toward production, curbing passive tendencies, and generally accepting, even with good humor, the responsibilities and sacrifices that come with the productive stance. By the same token, young mothers divest themselves, quite decisively, of the aggression that could put their vulnerable children at risk.

Having identified her own submerged aggression with the more flamboyant assertiveness of her husband, the wife further removes herself from those dangerous promptings by figuratively sending them out of the house, out to the perimeter, with him. Typically, across history, it is the women who have sung and cheered their garlanded warriors off to battle: Men go to war in large part because women respond amorously to uniforms. When men go off to battle, they are not only following some sexist bent of their own; instead, men become the exporters of aggression for both sexes. They carry on their swords and spears the sharp edge of their wives' hidden wrath, as well as their own. And even as they target their anger against other communities, they export dangerous aggression away from the vulnerable heart of their own settlements.

Further Reactions to the Parental Emergency: The Transformations of Narcissism

We have considered the realignments of "masculine" and "feminine" qualities, aggression and maternal love, in response to the parental emergency. But if we think of the parenting, conjugal family as a kind of organism, an extrauterine extension of the womb specially crafted to favor the psychological and physical growth of children, then it becomes clear that narcissism, the third great relational cement, must also be transformed and relocated. Clearly, the overarching familial structure could not be generated out of the union of a man and a woman unless both partners mutually renounced and transformed major portions of their own self-love and self-idealization. Indeed, unless parents were willing to forgo their narcissistic investments, their claims to omnisatisfaction and omnipotentiality, they would not undertake the necessary divestments—of sensual receptivity in the case of men and bridling aggression on the part of women—on which the family system is founded.

The nurturing organism of the family cannot come into being unless parents are willing to surrender many of their personal goals in favor of system goals, to become coacting parts—protective rind or nurturing core—of the family structure. If narcissistic goals were maintained unmodified by husbands and wives, the periphery would go undefended, aggression would be rampant within the sensitive domestic core, and nurturance would be reserved for the self rather than directed toward needful offspring. If narcissism is not transformed, redirected away from the self by both parents, the necessary family construction does not come into being, and children are gravely at risk: emotionally, because of their parents' unconcern, and physically, because of their parents' undeflected wrath.

While psychoanalytic theory mainly restricts the significant transformations of narcissism to the early developmental periods, implicating it chiefly in the formation of self and superego, any naturalistic review, unfettered by an exclusive bias toward early development, must convince us that new parenthood brings about one of the most potent transformations of narcissism in the entire life cycle. The difference between this and earlier conversions has to do with the fact that the parental transformation is not aimed at forging the adult self but is instead aimed at bringing about the psychological formation, the selfhood, of the offspring. New parenthood marks the point at which the conjugal couple routinely and, if things have gone as they should, automatically surrender a large piece of their narcissistic claims to personal omnipotentiality and immortality, conceding these instead to the child.[5] The result is the routine, unexamined heroism of parenting, which even renders mothers and fathers willing to die in their child's stead. Finally, parenthood does (or should) mark that point at which mothers and fathers revise their relationship to their own mortality. They may come to accept the natural staging of the life cycle, the inflexible order that dictates that they should predecease their children. Parents still fear death, perhaps more than ever, but now, more than their own demise, they fear the obscene possibility that the child might predecease the parent. Thus orthodox Jewish men introduce their first-born sons as their *kaddishle*, naming him with an affectionate diminutive as the one who will read the Kaddish, the prayer for the dead, over them at their burial. They thereby indicate their acceptance of their own finite life cycle, and even their pleasure in an order of things which holds that fathers, if all goes as it should, will die before their sons.

Indeed, the narcissistic transformations in response to the parental emergency are so profound that they force us to reconsider the ultimate significance of narcissism itself. Typically, we think of narcissism as being conserved for the individual, to provide emergency rations of self-love in the absence of external loving. And narcissism does indeed buffer individual self-esteem against the slings and arrows of daily existence. But its major function is not to protect the adult's *amour propre*; rather, sighting on narcissism from the parental perspective, we see that this tendency has been studied mainly in its immature and volatile forms, before it is recruited and stabilized to its true evolutionary and adaptive purpose: the preservation, by the parent, of the vulnerable child.

Entry into the Parental Emergency: Empirical Studies

Parenthood brings about major reorganizations of adult personality, and these should be particularly evident at two major transition points: the entry into and the exit from the state of active parenting. In this section, I present research evidence from various sources, all bearing on the psychological changes wrought in men and women by the onset of the parental emergency.

The formulations of Niles Newton (1973), based on cross-cultural data, help us to better understand the reasons for these crucial parental transformations. She argues that coitus, birth, and lactation—the three neurochemically regulated expressions of female sexuality—while they are vital to successful reproduction and child-rearing, are also strikingly vulnerable, prone to shut down in the face of outer threat. In order to proceed toward their reproductive goal, these activities all require external buffering and protection, most often provided by men. In this vein, Newton cites ethnographic descriptions of young mothers in South America, the Middle East, and China, all pointing to a standard pattern of maternal engrossment with the infant, in an intense bond that can persist through the first years of the child's life. At all these sites, the infant sleeps next to mother, is nursed at the first sign of restlessness, and nursing takes precedence over any competing activity. It is not surprising that, across most societies, young men protect wives who are so engrossed in the reproductive act and rendered vulnerable because of it.

An interesting if limited study done under my direction picks up the surgent changes set in train by first parenthood. Thus Gary Kupper (1975) interviewed and gave projective tests to newly parenting and nonparental married student couples, and found the predicted shift among the parental men toward disciplined careerism. Their goals, no longer inflated by narcissistically grandiose expectations, were more realistic and more attainable than those set by their nonparental age peers. Most interesting are the results from the intergroup projective test comparisons. The Heterosexual Conflict TAT card was shown to both the parental and the nonparental groups, and the resulting thematic distributions by parental status demonstrated differences in the subjective management of male aggression, along predicted lines. While most subjects see tension and discord in the scene, they vary by life stage in naming the sources of the couple's trouble. For the nonparental subjects, the scene is one of domestic squabbling: The young man and young woman are not sharply differentiated by their personal qualities or by the intensity of their aggressive feelings toward each other. In the stories from the parental group, strong sex distinctions appear, particularly in regard to the location and management of aggression: Tension no longer divides the couple; instead, these accounts feature extradomestic challenges—there is a war; there is a job that needs doing; another man has insulted the woman—and the aggressive response to these challenges is concentrated in the young man. His action is centrifugal: away from the woman, away from the domestic center, and toward the enemy or the opportunity that he glimpses on the periphery, beyond the actual card boundaries.

Conversely, any fearful concern over the young man's boldness is not found in him; instead, it is concentrated in the young woman figure, who tries to restrain the young man's reckless action. If he is all thrusting exuberance, then she is turned against such aggression and fears for its potential victims. In other words, as predicted by our model, for the parental group the young man figure has become a creature of the perimeter, whose job it is to concentrate the dangerous intradomestic aggression of both partners, to export it away from the vulnerable center, and to discharge it beyond the periphery against human or inhuman enemies, or against the agencies of impersonal nature.

But if parental informants see the young man as moving out of the domestic zone, to station himself aggressively on the perimeter, they also see women as moving back into the domestic zone. Consider the responses to the Farm/Family scene, a card that shows, in a bucolic setting, a man who plows the soil, a pregnant woman who leans against a tree, and a young woman who stands in the foreground, half turned toward the farmer and holding a book in her hand. Stories told in response to this card by normal subjects usually center on the actions and motives of the young woman. She is seen to be either moving away from the farm in pursuit of her own independence, or moving, usually motivated by guilt and filial concern for her family, back toward the domestic center, to be with and to help her parents. Our sample of nonparental subjects were prone to see the young woman as ambitious and centrifugal, but the parental subjects quite decisively placed her in the world of household and motherhood. Thus young parents move male figures away from the household, and by the same token they remove women away from the periphery, returning them to the center of the domestic world.

Katherine Ewing (1981), an anthropologist with psychoanalytic training, also probed beyond the behavioral level, exploring the subjective side of Pakistani mothers' reactions to their own parenthood. She found that their investment in maternal behavior is matched by an internal repudiation of their aggression, which is seen as dangerous to their offspring. In Ewing's experience, young mothers, afraid of the harm that unchecked aggression could cause their children, chastise themselves for their own uncontrolled outbursts of anger. The duty of the young mother is to submit to her mother-in-law and to control any anger over this servitude, for the sake of domestic harmony and for the sake of her children.

The predicted parental shifts are also noted for new American fathers. Thus Douglas Heath (1972) found that first-time American fathers are significantly more calculating and achievement-oriented than their nonparental age peers. By the same token, they are less emotive and less affectionate than married nonfathers in the same age range. In other words, young men move into parenthood by focusing on the role of provider and by damping out the kinds of eros that could interfere with that vital role. But these young men are not totally abandoning warm sentiment; rather, they are putting it on hold and giving over the feeling function to their wives. In effect, they become more stereotypically "masculine," not in the service of male chauvinism, but in the service of their children's need for a reliable provider: a father who will be able to leave the home when necessary, and without crippling nostalgia.

Studies of couples in early parenthood yield similar findings. Thus Richard Perloff and Michael Lamb (1980) found that, following the onset of parenthood, both partners in the marriage show increased sex-role stereotypy, while Elizabeth Menaghan (1975) claims that men without offspring are more passive than fathers of the same age, as well as possessing a higher capacity for intimacy. In effect, childless men, not yet stimulated by the parental emergency, show a "softer" personality style, in which feelings are given priority over unsentimental efficiency.

These Western-based studies tell us that the disciplined mobilization of the young man following marriage and especially parenthood is not exclusively a by-product of sexual training in preliterate traditional societies. Thomae (1962), citing studies of adulthood and aging among West Germen men, reports that younger men are impressed with the changes in their own personality that are consequent on parenthood: They find themselves "to be more stable, more of a homebody." George Gilder (1973) has also argued, from observations in American society, that men are tamed and "made social" only when they bond themselves to a woman and accept her procreative, parental purpose as their own. Until then, unmarried young men may be charming, dramatic, and attractive, but, except as warriors, they are socially surplus.

Standing Down from the Parental Emergency: The Emergence of Contra-sexual Potentials

We have seen how nature, society, and culture can meet, to support the development of sexually mature individuals into parental adults, those who will in their turn oversee the maturation of viable, pre-parental children. But as time passes, once-helpless children go on to demonstrate their viability by exercising for themselves the security-giving functions that had previously been exercised by others in their behalf. Children increasingly prove their capacity to provide physical security for themselves and others by demonstrating a marketable labor capacity, and by passing the endurance and fitness trials set by puberty rituals; they demonstrate their capacity to provide their own emotional security by developing a circle of tested friends, and by entering into intimate arrangements of some permanence. They emerge from their long period of dependency and helplessness, and they take over from the parents the executive functions that had been held in escrow for them.

With the phasing out of the parental emergency, fathers and mothers have less need to live within the altruistic conjugal mode. The child is safely launched and no longer needs to be protected, whether magically or realistically, by endowments of parental nar-cissism. In the postparental years, adults finish paying their species dues: They no longer have to meet the psychological tax that is levied on our species in compensation for human freedom from the pro-grammed rigidities of instinct. Having raised the next generation of viable and procreant children, the parents have earned the right to be again, at least in token ways, omnipotential. As a consequence, post-parental men and women can reclaim the sexual bimodality that was hitherto repressed and parceled out between husband and wife. Because the restoration of sexual bimodality will no longer interfere with proper parenting, senior men and women can reclaim, for themselves, those aspects of self that were once disowned inwardly, though lived out externally, vicariously, through the spouse. They can afford the luxury of elaborating the potentials and pleasures that they had to relinquish early on, in the service of their particular parental task.

In this later development, as we have already observed in the cross-cultural data, a significant sex-role turnover takes place, in that men begin to live out directly, to own as part of themselves, the accommodative, or Passive Mastery, qualities: sensuality, affiliation, and maternal tendencies—in effect, the "femininity" that was previously repressed in the service of productivity and lived out vicariously through the wife. By the same token, across societies, we see the opposite effect in women. As documented, they generally become more domineering, independent, unsentimental, and self-centered—asserting their own desires, particularly toward social dominance, rather than serving the emotional needs and development of others. Just as men in late middle life reclaim title to their denied "femininity," middle-aged women repossess the aggressive "masculinity" that they once lived out vicariously through their husbands. The consequence of this internal revolution, this shift in the politics of self, is that the sharp sex distinctions of earlier, parental adulthood break down, and each sex becomes to some degree what the other used to be. With their children grown, wives can tolerate and even enjoy the aggressiveness that once might have terrified their children and alienated their male providers; accordingly, recognizing such energy as a resource in themselves, they become less needful of and less admiring of male assertiveness. They take over some of the drive toward dominance that had previously been, almost exclusively, the hallmark and province of the male. By the same token, men are freed up to recapture the latent duality of their own nature: They become more hedonic, more dependent, and also more irritable. That is, they live out again the extremes of their psyche, tolerable now that their aggression is biologically reduced, taking on a peevish rather than a murderous form. In effect, besides claiming some of the cast-off parts of the feminine psyche, men also reclaim title to the full spectrum of affects that their now grown-up children have in their turn abandoned. They can be childish: needful, querulous, indignant, or all of the above. In sum, as the parenting establishment is demobilized, the senior individuals take back into themselves the qualities that had been portioned out into the various parts and agencies of the supraindividual family structure: Men take back the sensuality and tenderness that they once left behind at the domestic center, while women take back the aggression that they, through their husbands, had once shipped off to form and defend the communal periphery.

Finally, both parents reclaim the quotient of narcissism that had been transformed to the service of children and that had constituted the cement of the child-centered, child-tending family. As that social structure is dismantled, narcissism again becomes available to the postparental self. Depending on the social circumstances under which it emerges, that fund of narcissism can become the basis of new self-idealizations (as well as new vulnerabilities), or it can become recruited to the formation of new, more extensive social bonds. These can tie the generations to each other, or (as will be discussed in chapter 9) they can link the community to the gods.

The Exit from Active Parenthood: Empirical Studies

In recent years, research findings have documented the relationship between parental status and late-blooming features of personality. While a number of researchers have reported sex-role changes in the postparental years that conform to predictions from my model, I will consider in detail only those that control for the effects of chronological age, thereby highlighting the psychological effects directly traceable to life stage, to the exit from active parenting.

As with the studies that covered the entry into the parental state, I will review first those studies that trace postparental changes at the more surgent levels of personality.

The Druze Case: A Cross-cultural Test of the Parental Model

My Druze research provides an opportunity to test the universality of the relationship between the parental stage and the male psychological changes of later life. As we have seen, in later life the focused, "phallic" organization of masculine personality tends to shut down in favor of a more diffuse spectrum of erotic possibilities, including those that were salient in the earliest years of life. For example, we saw earlier that oral-erotic interests, evidenced by the pleasure in eating, show the same prominence in later life, across the Navajo and Druze societies, that they presumably showed in early childhood. To test the possibility that parental status has an effect, independent of age, in determining the variations in orality (and, by extension, in passive-dependency), a scale was developed to quantify the levels of parental involvement among Druze men. On the basis of interview data, Druze informants were classified as to whether they were actively involved in parenting, with young children in the home; were phasing out of active parenting; or were fully into the postparental period, with all children "launched." Figure 8.1 shows the distribution of orality scores by age and by stage of parenting, and indicates the powerful effect that male parenthood plays in determining the oral, passive toning of Druze personality. Thus, for any age, men who are still actively parenting show lower mean orality scores than the post-launch men. While the level of orality remains relatively constant across age groups for those men who still maintain dependent children in their households, these same orality scores go up, dramatically, across all age groups, for the Druze "empty-nest" men. Again, it is not chronological age that brings on the psychological signs of aging; in the more inner and subjective sense, aging comes about when men demobilize, standing down from the parental emergency.

DRUZE EGO MASTERY SCORES. We get similar results when we consider age and parental status as independent variables, and distribute the TAT-derived ego mastery scores of Druze males against these dimensions. These TAT ratings express in numerical terms the proportions of Active, Passive, and Magical Mastery responses in the individual test records, the hypothesis being that parental stage would prove more powerful than chronological age in accounting for the variation in ego mastery scores. Specifically, I believed that the cessation of active parenthood (marked by the absence of dependent children in the subject's household) would predict, more significantly than age per se, to reduced scores on Active Mastery and to elevated scores on Passive and Magical Mastery modalities. As figure 8.2 indicates, these expectations were borne out to a significant degree: Analysis of variance tests reveal that prelaunch, still-parenting Druze men are higher on Active Mastery than their postparental age peers.

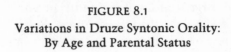

FIGURE 8.1
Variations in Druze Syntonic Orality:
By Age and Parental Status

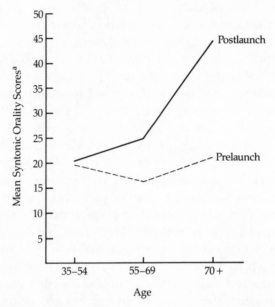

^a The mean difference between the prelaunch and postlaunch scores is statistically significant at the .065 level by Analysis of Variance (F test).

Most significant is the finding that men in the 55-and-over age range who still parent score even higher on Active Mastery than younger postparental men in the age range 35–54. Though there are independent parental stage and age effects, the stage effects are most significant, by statistical test.

The predictions from the parental model are also borne out in regard to Passive and Magical Mastery, though there is an independent age effect registered in the sequencing of these mastery styles. As figure 8.3 and figure 8.4 indicate, most postparental Druze men, re-gardless of age, show reductions in Active Mastery; but younger post-parental men shift from Active into Passive Mastery, while older postparental men, as they exit from Active Mastery, shift directly into Magical Mastery, bypassing the intervening Passive Mastery stage. Again, the major thesis of the study is borne out, though some-what qualified in form. The postparental transition does release, in men, a renewed narcissistic demand for the kinds of sensory pleasures

FIGURE 8.2

Variations in the Percentage of Active Mastery
TAT Responses among Druze Men, by Age and Parental Status

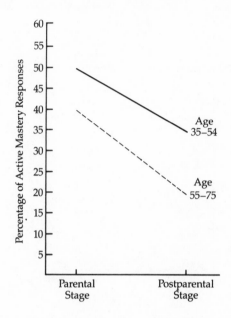

and egocentric thinking that they had previously muted in the service of their parental assignment. However, as an independent factor, age determines the choice of postparental modalities—toward dependency and diffuse sensuality (Passive Mastery) for younger men and toward primitive thinking (Magical Mastery) for older men.[6]

Three American researchers who recently undertook similarly explicit tests of the parental model generally confirm these impressions. Comparing late fathers (both the continuing and the late-starting) to a matched sample of postparental men, Linda O'Connel (1980) found that the former group showed the typical signs of the younger, Active Mastery stance on most of her instruments. The TAT protocols of the late fathers showed a higher percentage of Active Mastery themes, their relationships with their wives were marked by clear, stereotypic sex-role distinctions, and they tended to give priority to work and achievement over sentiment in their daily lives.

Donna Ripley (1984) carried out a dissertation study of sixty mid-western male industrial workers sorted into four groups: younger fathers, younger "post-fathers," older fathers, and older "post-fathers"

FIGURE 8.3

Variations in the Percentage of Passive Mastery TAT Responses
among Druze Men, by Age and Parental Status

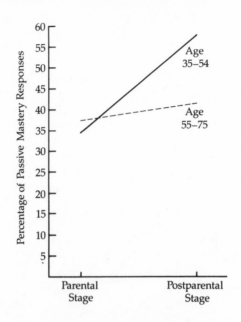

(for both postparental groups, the last child has been launched). These
men were given projective tests and interviews from which an overall
"gender" score—ratings that reflected the degree of Active Mastery
orientation—could be derived. Supporting the parental imperative
hypothesis, Ripley finds that Gender Index Scores reflecting an active,
"masculine" stance are highest for parental men, regardless of age,
and lowest for the older and younger postparental men.

Kathryn Cooper (1987) carried out an equivalent TAT study of
employed female teachers, matched as to age and distinguished by
parental status; half were prelaunch, and half were postlaunch. The
prediction, that postlaunch women would show the signs of surgent
change toward more aggressive, confident, and even masculine postures, was borne out (see figure 8.5) for all but one TAT card, at statistically significant levels. As a result of their liberation, the postlaunch women, far from grieving over the emptied nest, now fill the
TAT figures with energy and single-minded purpose. For the prelaunch
matrons, the young woman of the Farm/Family card is torn between
the wish to leave home on her own career mission and guilty concern
for the hardworking parents who would be left behind; but for the

FIGURE 8.4

Variations in the Percentage of Magical Mastery TAT Responses
among Druze Men, by Age and Parental Status

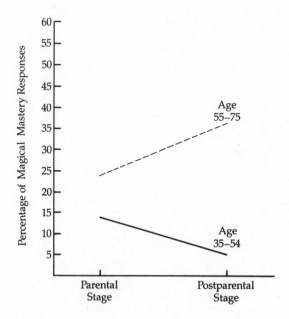

postlaunch women, the same girl takes off on some legitimate career mission without a backward look. The shifts in other cards are along the same "self-actualizing" axis.

Although they will not admit these assertions publicly for some years, at this surgent level of communication the postlaunch women are telling us that they are finished with overseeing the growth of others and are entering the phase of self-development, of tending their own growth. This personal flowering may take place in the context of a caring profession, such as clinical psychology, or at the executive levels of the extended family, but the emphasis in midlife is on the professional and careerist aspects of the role, rather than on the provision of care for its own sake.

Using the TAT as a seismograph, to pick up early rumblings in the guts of the psyche, Cooper has explored the surgent events set in train by postparenthood. Using more "public" methods, aimed at getting at more finished and sculpted outcomes, Suzanne Galler (1977) studied professional women, much like the women in Cooper's sample, the difference being that Galler's subjects are already some years into their "self-launching." Not surprisingly, Galler found that female

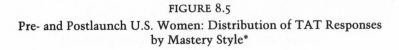

FIGURE 8.5

Pre- and Postlaunch U.S. Women: Distribution of TAT Responses
by Mastery Style*

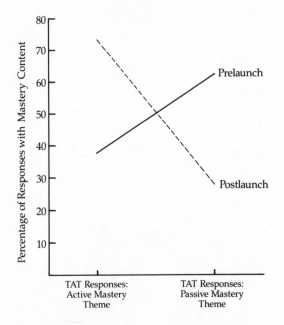

* I am indebted to Kathryn Cooper, Ph.D., for the research data on which this figure is based.

midlife returnees to professional work had always been achievement-oriented, but that their dormant ambitions had been quickened by the phasing out of active parenthood:

> Women in the professional student group describe themselves as having kept their personal ambitions in the background for many years, supporting, instead, the career aspirations of spouses who are themselves professional or academicians, as well as investing themselves in child rearing and adjunctive community activities. The "it's my turn now" attitudes verbalized by many of these returnees express a self-assertive rather than an other-directed attitude, an acceptance of their current limited interest in domesticity.

Using the recall of early memories as a projective technique, Harvey Peskin and Norman Livson (1981) conducted a particularly significant study, carefully sorting out surgent psychological changes attributable to the parenting stage per se, as distinct from those attributable to chronological age (and the social expectations that pertain to age). Inviting parental and older postparental adults to reminisce about their adolescent years, they found, as expected, that current parental status apparently shaped the recall of earlier experience. Regardless of age, actively parenting women tended to recapture memories clustered around nurturant and sentimental themes, while "empty-nest" women, again regardless of age, were more likely to stress memories that were charged with themes of power, dominance, and autonomy. Thus the memories of actively parenting 40-year-old women were like those of actively parenting 30-year-olds, but unlike the memories of their postparental age peers. Similar effects were observed for men. Actively parenting men remember their adolescent years as a time of competition and training for leadership, while postparental men in the same age range centered their memories of adolescence around milder, more affiliative themes: campfire sings, rather than drag races. Again, the memories of parentally involved 40-year-old men were like memories of 30-year-old parents, and unlike those of postparental peers from the same age cohort.

Studying the more proactive and sculpted aspects of women's roles, the anthropologist Sylvia Vatuk (1975) studied the Rajput Raya, finding again that parental stage counts more than age in demarcating the divisions of their later lives. Thus women enter the age grade of "old age," regardless of their actual years, when their children, particularly their sons, get married. However, the Raya women's transition into old age brings with it social gains rather than losses. Her son's marriage may "age" her in the social sense, but it also brings her daughters-in-law who are pledged to her service; in particular, they do the "inside" work, while the older woman now occupies herself with "outside" work, on the interface between the family and the larger community. Vacating the classic woman's role at the domestic center, the postparental Rajput women move, like younger men, to their society's perimeter.

Also working with the more social fabric of personality, Shirley Feldman, Zeynep Biringen, and Sharon Nash (1981) used the BEM Sex Role Inventory (BSRI) to study fluctuations in the sexual orientations of California men and women across the stages of life from adolescence to grandparenthood. While the BSRI, as a self-report questionnaire, is not designed to tap the surgent, unconscious attitudes and motives that are perhaps most important in determining sexual orientation, and while it could be particularly vulnerable to respondents' "social desirability" concerns—their wish to appear appropriately "nonsexist" to liberal social scientists—the authors nonetheless report positive findings:

> [The] speculation that instrumental and expressive attributes are better understood as specific to stage of life rather than as traits affixed to one sex or the other receive some support from our data. . . . The largest sex differences (on expressive factors) occur during the active parenting stages when in traditional families they serve to promote effective primary child care. The sexes describe themselves more similarly by the time that all their children have grown up and left home. . . . With the alleviation of the role constraints of active parenting, both sexes perceive themselves as having more cross-sex-typed assets yet do not feel they are any less masculine as men or feminine as women.

What is most striking here is that the young parents in the United States, despite the massive pro-androgyny rhetoric to which their generation has been subjected, nevertheless demonstrate the sex-role differences during the actively parenting years that are predictable from our model: the same polarizations that occur on schedule in more traditionally "sexist" societies.

In the more general or species sense, the studies reviewed in this chapter suggest that the psychological qualities called "masculine" and "feminine" are more tied to the seasons of parenthood than they are to biological sex. It is the balance and scheduling of these qualities (rather than their exclusive possession) that distinguishes men from women. Men comprise the sex that is univocally masculine before reclaiming a feminine component; and the female sex reverses this staging. But despite these life-span changes in the mix of masculine and feminine qualities, both sexes are alike in that men and women maintain their core sexual identities over the years. As in other areas of life, the socially-derived names, such as "man" or "woman," have more endurance and are less changeable than our ever-shifting biological nature.

The reported findings also suggest that genital primacy is not (as the psychoanalysts would claim) the end point of sexual development. In the genital stage of erotic development, the capacity for pleasure, hitherto distributed over the entire body and its organs, is claimed more exclusively by the genitals, serving the purpose of species pro-creation as well as individual pleasure. The libido, the capacity for pleasure, is fettered, restricted to the genitals and to the occasional, fleeting ecstasy of the orgasm. During adulthood the utilitarian, species-preserving principle of the erotic life dominates the pleasure

212

principle. Thus, contrary to Freudian doctrine, genital primacy is not the final goal of eros but a kind of transient deviation in its course. It is a grudging response to the parental emergency that is undone when the crisis passes. The final development of eros comes about in the postparental period—the capacity for genital pleasure is retained, but the receptive pleasures of the mouth, eyes, and skin are regained. Given reasonable health, a more complex, variegated, and interesting consolidation of the erotic life is finally attainable in the later years.

Finally, those studies that point to a parental armature of psychological change in the later years, though far from complete, do suggest a productive direction for life-span developmental psychology and a hopeful redirection for geropsychology as a whole. For example, they provide refreshing alternatives to the catastrophic view of the aging process, which holds that all later-life changes, in men and women, are ultimately last-ditch adaptations to imposed loss and inevitable depletion.

Our developmental model, as reviewed thus far, holds that the ending of parental restraints liberates unused, blunted potentials, and that these become available for new satisfactions and uses. In the final chapters, I look more closely at the ways in which these surgent energies are transformed into new psychological structures for old people: the executive capacities that serve self, society, and species in later life.

9

ELDERS AS EMERITUS PARENTS: THE STRONG FACE OF AGING

They came. They brought the ones who had been killed by the white people. My aunts were with me. My mother, my father, my aunts, held me and went with me. I came there; I was pregnant. They would not let me see him, my husband. Only my mother saw him. She told me. It was not good. . . . So they buried them in the graveyard, just before sunset. . . .

My grandfather took care of me. "It is very dangerous; you must fast. You must drink medicine. You must vomit. It is very dangerous. No one may touch you. It is very dangerous, you must fast. No one must touch you. You must stay alone. You must sit alone in the corner. Only your little boy may hold you. No one must touch you." Grandfather gathered medicine for me. This he soaked. He mixed it in a fine bowl. He brewed medicine. "This you will drink. You will vomit," he said to me. I was very wretched. This was very dangerous. When it was still early, when the sun had not yet risen, my grandfather took me far away. We scattered prayermeal. Here in the left hand I had black prayermeal, and here the right kind of prayermeal. When we had gone far I passed it four times over my head and scattered it. One should not speak.
—"A Zuñi Woman Mourns for Her Husband,"
quoted by RUTH BUNZEL (1930)

We saw in chapters 6 and 7 that the traditional extended family setting provides a sponsoring milieu for the older woman's maturation, a setting in which the genotypic, surgent potentials released by the ending of parenthood are crafted toward their phenotypic outcomes: the various forms of kin-tending matriarchy. In this chapter, I argue that equally late-blooming potentials in older men find their way to matured outcomes in and through the sponsoring environment of the patriarch, the realm of traditional culture and its associated institutions.

The Human Male: Surviving to Elderhood

We start by recognizing that we are dealing with a unique human event: Older primate males do not generally survive into the post-parental years and beyond the peak years of muscular strength. In the case of female elders, their social assignment, as matriarch, is a continuation of the kin-tending role that, as we have seen, had already evolved among the subhuman primates. But there is no discernible postparental role for the older male ape. When he loses physical strength, he falls out of the dominance hierarchy and, as a consequence, loses his sexual access to females. Seemingly depressed, he soon drops out of the troop. He either dies or wanders off in search of a band in which his remaining strength is still equal to the competition. In either event, he is lost to his troop, and his accumulated experience goes with him.

It is only when we cross the ape-human boundary that we encounter the uniquely human condition of male gerontocracy. Despite the older man's inner movement away from aggressive domination, the ethnographic literature makes it clear that—at least in preliterate, traditional societies—his official authority tends to increase. Simmons (1945), for example, finds that male gerontocracy is almost universally the rule in the folk, tradition-directed society. Specifically, his review of seventy-one tribes, from all parts of the world, shows that older men were chiefs in fifty-six of the sample societies, that they could choose to hold office in eight more, and that in no case were they deposed on account of age.' Humans then are the only primate species in which the older male, despite his loss of physical strength, can remain in the human band and find alternate dominance hierarchies wherein status depends on qualities other than brute power, cunning, and ferocity.

In the human case, institutions that are as unique to our species as male gerontocracy itself may underwrite the physical, psychological, and social survival of the aging man, at least until he can gain his own secure place as a sacrosanct male elder. The incest taboo, for example, guarantees him sexual access to at least one female, his wife, for whom he does not have to fight. The young men, including his sons, will not usually challenge his sexual rights to a "mother figure." (As we have noted, young men sometimes replace their fathers as coleaders, with the mother, of the family; but they do not fight the father for these powers. Inheriting what he abandons, they do not take over his sexual prerogatives with the mother, they do not drive him from the community, and they continue to accept his leadership in spiritual if not in practical matters.)

Thus, except when he becomes a widower the older man is usually spared depression, the common fate of unmated older males. From this evolutionary perspective we can also understand the utility for the species of later-life male androgyny, the resurgence of sexual bimodality in older men. We saw that for the older female, subhuman and human, eruptive aggression is inducted to species service via the kin-tender's role; the older man's sexual bimodality plays its own crucial part by exempting him from competition with younger, physically stronger men. In our own time, we see that controversial politicians lose their capacity to create and challenge opponents as their gray hairs thicken. The deposed Nixon is reborn in his seventies as an "elder statesman," a status that he now shares with Barry Goldwater, and the aged Ronald Reagan, our "Teflon" president, has been (at least until recently) spared the invective that usually batters the Chief Executive.

Apparently, the androgynous older man is above the battle; he does not stir the sense of threat in young men, and there is no honor and much guilt to be gained from killing him. Thus enabled to hold on to their women and certain versions of social power, the older traditional males do not die as their muscles weaken; instead, they keep title to their wives and honors, stay in place, add their accrued experience to the general knowledge store, and—in ways to be reviewed in this chapter—build culture. This species achievement, of culture via male gerontocracy, is recapitulated in the ontogeny of male development in later years. Under proper conditions, "normal" old men disengage from their life in society while enhancing their life in culture, developing themselves as culture-tenders.

Male Gerontocracy and the Folk Society

The incest taboo and elderly androgyny may preserve the older man long enough to develop his new function, as a senior tender of culture, but what special qualifications does he bring to his gerontocratic position? How does he enhance his special executive status, particularly in view of the qualifying evidence presented earlier, showing the emergence of more passive, narcissistic, counterdominant trends in the older man? How is it that despite the outward stigma of their age, as well as their own inner "maternalism," older men manage to garner the kinds of respect that make for male patriarchy and gerontocracy?

We may penetrate this seeming paradox by considering the special roles and social functions of older men, as well as the ways that these contribute to their prestige. We have already observed, in the chapter on male age grading, that younger and older men are routinely shunted toward different careers. Younger men are tracked mainly toward activities that are founded on outward-turned aggression—warrior, entrepreneur—activities that are checked only by the opposition of nature or other men, while older men track toward activities founded on self-control, inhibition, and self-effacement.

A substantial number of ethnographers have identified the kinds of prestigious social roles that give expression to these milder, controlled tendencies of the older man. As they report (and their accounts accord with my observations of the Navajo, Highland Maya, and Druze, as laid out in chapter 2), when societies maintain traditional religious institutions, older men lead the congregation. The correlational study by Timothy Sheehan (1976) of the relationship between societal type and senior prestige reveals that, for a sample of forty-seven preliterate groups, a settled, nonnomadic, relatively isolated, and traditional society provides the best ecology for the aged. Under these conditions, there is an intellectual and cultural life, there is folklore to be guarded and handed down by seniors, and there is religious thought, cosmic ceremony, and *rites de passage* through which they are linked to the gods.

The general rule seems to be that older persons are less (rather than more) apt to control resources that have to do with pragmatic, economic production; but by the same token, their grip over the ritual sources of sacred power increases with age.

The slowly accruing cross-cultural evidence suggests that it is not

217

their store of material possessions but their special association with the sacred systems of society that is the true key to the traditional elder's prestige.[2]

Passive Mastery, Power, and Patriarchy

Paradoxically, this vital linkage between sacred institutions and the traditional elder is guaranteed by the same qualities that lead to social disengagement and even psychopathology under secular conditions. Again, the major paradox of primitive gerontocracy is that older men gain spiritual status not because of a need to dominate, but because of the same passivity, even the self-abasement, that moves toward prominence—and that preserves them from bruising combat—in their postparental years.[3]

A central feature of the traditional society is that potentially stig-matized qualities, such as passivity, can attract great power to old men; this paradox is in turn founded on a basic division between two aspects of the traditional culture. It is not only divided between sacred and secular aspects, but it also maintains two conceptions of power, pragmatic and sacred, each one particularly suited to a distinct season of the life cycle. Pragmatic power is tangible, available to the senses, and lodged in the muscles. It is the energy of the sexual body, the energy of the young. But the ultimate life-sustaining and life-destroying sacred powers are split off from the physical body and the mundane community; they are lodged in the various domains of su-pernatural power that surround the immediate, experienced com-munity. In the folk mind, the pragmatic, daily community is an island in a sea of *tabu* power that beats up on the shores of the known world but that also stretches away into the disquieting reaches of the un-known. The world-sustaining mana, or sacred energy, has many sources, and these vary by culture, but folk societies generally main-tain the idea that the prosaic world is kept "real" and vital by power imported from those agencies that penetrate into the daily world but that also extend into some mythic domain outside of it. Wherever this world view is institutionalized, anointed figures are required to live on the dangerous border between the mundane and supernatural worlds, in order to "attract" the benevolent aspect of supernatural power, to contain it, to neutralize it, and to make it available to the human community and its ecosystems. In the preliterate mind, these intermediaries are typically the bridgeheads of the unordinary domain within the ordinary world; they freight in the vital power that literally keeps the mundane world alive. A number of agents can fit this special power niche: the dead (particularly the enemy dead), the madman, and the aged (particularly aged men). All have in common these qualities: They are unordinary, of this world and yet unfamiliar, and they are passive, broken by fate, by the gods, or by time. Empty of pragmatic power, they become reciprocals of or conduits to *tabu* power, the vital energy of the gods. Thus Róheim (1930) has cited a large body of data from preliterate societies, all documenting the point that the "transformer," or power-bringer, role is enacted through submissive and even masochistic behavior before the gods; across cultures, the gods are mainly seen to give power to those men who approach them humbly, sometimes even in the guise of women.

For example, in the Siberian cultures, the theme of masochistic femininity as the route to power was made quite explicit. According to Róheim, the Chukchi shaman invoked the spirits in the guise of a woman: He wore a woman's cloak, parted his hair in woman's fashion, and wore two iron circlets, representing breasts, on his shirt. When possessed by the spirits, he would writhe on the ground, miming the motions of a woman in orgasm. However, at the height of his ecstasy, he would cry out in a deep, resonant voice that he was the mighty bull of the earth, set above all beings by his all-powerful master. By playing the woman before the gods, by being passively possessed by their power, he had acquired that same power and had become mightier than any man. While the gods appear more willing to give their potency to men who approach them as yearning women rather than as promethean rivals, the basic equation holds: He who can endure, however passively, the awful god power without being destroyed by it will become the temporary vessel of that power.[4]

Accumulating the gods' *tabu* power for social purposes is not a task for promethean young men. The necessary passivity is alien to them; their boldness would offend the gods and bring disaster on the community and on themselves. To avoid this fate, the power brokers who man the sacred perimeter of the community must, in their own nature, be metaphors of the "good" power that they would attract. Accordingly, in the traditional and religious community, it is the postparental older men, cleansed of sex and aggression, humble in bearing and even maternal, who can beseech the nurturant influences of the gods, without offending the divinity. In effect, older men turn the feminine aspect, which was once lived out vicariously through their wives, to the love and service of the gods. In any event, the sweet, "maternal" old man becomes a vessel for the nutritive, life-restoring "good power" of the gods (while the mean and angry old woman becomes, as we have seen, a metaphor of bad supernatural power, and is therefore suspected of witchcraft).

This transformed, even uncanny, quality of the old is caught in a profound observation by de Beauvoir (1972), who points out that the aged, in that they can represent the stranger, the "other," are always at risk, even under the most benign cultural circumstances. Universally, whether he appears in the guise of the enemy, the corpse, the madman, or the elder, the stranger calls up mingled feelings, of revulsion and of awe; the traditional society, because it puts the old man in the power stream, emphasizes the aspect of awe, and lends him a magic aura that overcomes the tendency toward repulsion.

Mana and Aging: The Old Man as Hero

There exists, then, a generational rule of some universality that compensates the traditional aged for their "stranger" potential and for their losses of physical power, by the acquisition of supernatural power. As detailed in chapter 4, young men kill with edged weapons, but older men can kill with a curse. Thus Lucien Lévy-Bruhl (1928) observed that, in the preliterate folk society, the old man is "encircled by a kind of mystic halo," an essence so pervasive that his body parts and even his excrement can become the residence of *tabu* power. Even Simmons (1945) moves away from his usual stance of rationalism in this observation:

> It was not strength or brawn alone that won in battle or staved off bad luck or healed the dreaded disease; it was a special power, mysterious and most potent in the hands of old men and old women who have survived all these dangers. . . . Not *all* magicians were old, but superannuation and the supernatural were very commonly and closely linked.

In illustration, Simmons reports many examples of the older man's awesome *tabu* powers, the most striking coming from the Hottentot of Africa, where the old men initiate young men, who have passed their early life among women, into manhood. The climax of the rite comes when the old man urinates on the candidate, who receives the urine with joy, rubbing it vigorously into his skin. His old sponsor then tells the candidate that he will increase and multiply, and that his beard will soon grow. Clearly, in this case, even the urine of the old man has heroic power, the mana of the patriarchal phallus through which it passed. In the most concrete sense, it "marinates" the young man with the powers of the old man, thus bringing the lad in his turn to manhood.[5] Here we see, unequivocally, the strong face of aging, the face that is hidden from us in our own secular, contra-gerontic society.

In sum, whether by virtue of their special weakness or their special strength, the aged are elected. While young men live on the physical perimeter of the community, to contain and harvest the forces of ordinary nature, the old men retreat physically to the interior, domestic zone. Having established a secure home base, they then move out to the spiritual perimeter, there to fend off the bad power and to harvest the good power of the gods. As in most developmental sequences, the seeming withdrawal is the precondition for a later advance—an imaginative leap outward, to the supernaturals.

To repeat, older men may lose the qualities of the warrior, but as these phase out they reveal an understructure of hitherto-hidden cognitive and affectional potentials. "Hardened" into fixed capacities, these are recruited to the service of self, society, and species in the later years. As men give up the ways of the warrior and their stations on the community's perimeter, they move back toward the domestic world, there to rekindle qualities of sensuality, emotionality, and mildness repressed during their days of fighting and fathering. Rather than restlessly seeking and provoking change on the perimeter, they seek constancy in terms of place, person, and nutriment at the protected center. However, in their minds they still reach out, beyond the familiar neighborhood, to invest the supernaturals, the ultimate vessels of unwavering wisdom and nurture. Older men who have fallen back from the physical perimeter subsequently, as masters of ritual, move out to the spiritual perimeter, to confront the powerful and empowering gods. Older men discover in the supernaturals the strength that they no longer find within themselves; they use prayer, a passive-dependent modality, to beseech, for themselves and for the people, their fire from the gods.

Inevitably, as wardens of *tabu* power, the aged also become *tabu*;

social prestige follows sacred power. Emptied of pragmatic power, the older man becomes a fit vessel for sacred power, and his blessing can uplift and his curse can blast with a force, borrowed from the gods, that was not available to his younger self, and is not available to his son.

Gerontocracy and Religion among the Druze

These developmental transformations, of passivity into power, are dramatized by older Druze men as they enter into and maintain the religious institutions that define their society. In the Druze case, there is a coordination between the ego, its eruptive potentials, and the facilitating psychosocial ambience created by the Druze religion, which transforms these potentials into sculpted executive capacities for the service of the individual and for the Druze people as a whole. We have already seen that the older Druze man gives up many indulgences, in thought and in deed, to enter the ranks of the *aqil*, "the knowing ones," the privileged keepers of the Druze faith. In exchange for his sacrifices, the *aqil* becomes a figure of vast dignity, deferred to by the young, an enduring rock of the community.

But if the passive leanings that emerge openly in the TAT protocols of elderly Druze are universal in scope, we can expect them to also be powerful; they should have a peremptory effect on behavior. Yet, as noted earlier, the public behavior of the older *aqil* seems to be completely ordered by his active role in local society, not by some specieswide undertow toward passivity. When we look at the conventional behavior of the religious man, we find him going busily from place to place, from one ceremonial visit to another, involved in praying or receiving guests with elaborate hospitality. In his social relations, he is definitely not submissive but dogmatic and even dictatorial, laying down the law to his younger relations. Within the public framework, the older *aqil* seems to behave in an actively coping rather than a passive fashion.

When we ask the *aqil* what he does as a religious man, he describes an energetic life. However, when we ask him about the meaning that the religious life holds for him, his subjective relation to the religious life and to God, we get a different picture. It is at this level that the passive yearnings inferred from the TAT seem to make their appearance, in such a way that the *aqil* resembles his overtly passive age-peers in other societies. When they talk about their relationship to Allah, fierce, patriarchal old Druze, who fought the French Foreign Legion and who still rule their grown sons, adopt completely the posture and the tone of the passive, self-effacing supplicant: "Allah is all and I am nothing; I live only in his will, and by his will. . . . I do not complain about my illness, because this is from God, and to complain about my illness is to question the will of God." These older men are not playing back the prescription for conventional religious behavior. When we ask the few young *aqil* about the personal meaning of their religious life, they tell us that their task is to seek out sinners and correct their ways. Through this action they say, they make the village acceptable in the eyes of God. In effect, the younger *aqil* are sheriffs for God. They clean up Dodge City; their relationship to Allah is mediated by their aggressive, policing action in his service.

It is only among the old *aqil* that we get the sense of a direct and personal relation to Allah, mediated not by work, but by supplication and prayer. Young *aqil* change the village; old men make *themselves* acceptable to Allah, who is for them an intensely felt and loving presence. As they talk of God their eyes shine, and the voices of old patriarchs tremble with emotion. They become not unlike the stereotypically submissive woman who speaks fearfully and yearningly of her master. There are no rules in Druze society that tell young *aqil* to have a relation to Allah that is centered on their own action or that direct older men to a relationship centered instead on the power and actions of God. Clearly there is a range of permissible postures toward Allah available to the Druze, regardless of age, and each age cohort finds certain stances within this range more congenial than others. The younger *aqil* discover a relationship to Allah that is in conformity with the principles of Active Mastery, while older *aqil* enact the themes of Passive Mastery; these age preferences reflect intrinsic rather than social coercion: nature, again, rather than nurture. Evidently the religious role allows the passive strivings noted in the older man's TAT to find their dramatic though normative expression, their sculpted outlet. The forms of social nurture decide the mode of expression, while nature provides the driving, surgent motives. Predictably, these motives announce themselves most directly through the projective system of the Druze religion.

The Druze case illustrates the relatively seamless fit that can exist between particular roles and developing gero-psychic potentials in the traditional community.[6] As we have seen, the older Druze shares with his age peers in other societies the tendency toward what might be called the normal androgyny of later life. However, he does not need to make some final and conflictful choice between active and passive, "masculine" and "feminine" relational styles. Through the psychological tactic of splitting, the *aqil* can keep title to both: The traditional religious sector of society provides a particular psychosocial ecology, a projective system in which he can live out passive and even "feminine" strivings, while he continues to domineer his sons in the home (where he is known as the "Allah of the house"). In effect, by being self-negating toward God, the *aqil* lives out the ancient dialectic of power. His self-abasement in the private setting of the hilweh, the prayer house, gives him the endowment of power, experienced as self-esteem, that he requires for his leadership role within the community. Assured of his psychic supplies from extracommunal, divine sources, he can take strong positions in the community, even if these sometimes earn him resentment rather than love.

In short, the role of *aqil* requires and gives definition to those psychic potentials that are released by the older man's withdrawal from the active tasks of parenthood and production. The yearnings that men in secular societies might experience only obliquely—through their denial or in the form of neurotic symptoms—the old traditional Druze experiences in two ways: as his worshipful linkage to God, and as his freedom from the judgments and concerns of mere men.

Disengagement and Reengagement

The theory of later-life disengagement put forward by Elaine Cumming and William Henry (1961) in essence proposes a mutual decoupling, whereby the agencies of society withdraw their attention from the aged and the aged withdraw from society's normative restraints, to become more idiosyncratic but also more "liberated." Though the theory was developed exclusively from urban studies in the United States and was not tested cross-culturally, the authors claim that disengagement is both mandatory and universal. Disengagement from internalized social control is seen as a developmental event; the older person who sets himself to oppose this dictate of nature is fighting a losing battle and may even do so at his peril.

Cumming and Henry partly justify the case for a developmental underpinning to disengagement by relating this process to another presumably developmental event: the later-life emergence, in men, of passive ego states. Since the age trend toward TAT expressions of passivity matched the age trend on other social barometers toward disengagement in the study population, Cumming and Henry concluded that the two trends were linked into one development event, in such a way that the increased passivity of later life represented the inner, subjective corollary of the total disengagement process.

However, the case of the Druze *aqil* indicates that disengagement need not be compulsory, and it particularly demonstrates that accommodative ego states are not inextricably tied to disengagement. Quite the contrary: In the Druze case, and probably in the case of other folk-traditional societies with a strong religious bent, the so-called passivity of the older man can be the central, necessary component of his engagement with age-appropriate social roles, traditions, and associated normative controls. We see, for example, that the older Druze *aqil* switches his allegiance from the norms that govern the productive and secular-productive life to those that govern the traditional and moral life, but in this transition he does not stray from the influence of normative controls as such. Instead, they gain increased influence over him (if anything, his accommodative tendencies make him all the more sensitive to social requirements). The older Druze may detach his interest and allegiance from those social codes that are no longer congenial to his passive needs, but he certainly does not detach himself from social norms as such. Instead, he links himself subjectively to the religious dimension of his society, and in so doing plays out the dominant motif of Passive Mastery: the need to be in personal touch with a powerful, benevolent, and productive agent. He relinquishes his own productivity but not productivity per se. Instead of being the center of enterprise, he becomes the bridge between the community and the productive, life-sustaining potencies of Allah. The old *aqil* now carries forward the spiritual rather than the material work of the community. Guided by needs and sensibilities that reflect his emerging passivity, the older Druze transits from one normative order to another within society; in that transition, he becomes the instrument and the exemplar of the traditional moral order that he has adopted and that has adopted him.

What is true for the Druze is also in general true for the men of other tradition-oriented societies. As we already learned from the reports of Simmons (1945), the traditional elders of preliterate groups do not usually disengage from the social order and its normative prescriptions; on the contrary, they often become the interpreters and administrators of the moral sector of society. They become the norm bearers. The disengagement that Cumming and Henry found in our society is the exception rather than the rule. It is only the first step in a total process of transition and reengagement, a process that can reach its natural terminus in a traditional society but that is interrupted or aborted in a secular society. It may well be that disengagement is an artifact only of secular society, which does not offer the old man a traditional moral order to which he can relate once he has decoupled from those social norms that regulate the parental and productive life periods.

In sum, it is the elder's movement toward Passive and Magical Mastery that appears to be universal, not the movement toward disengagement. The inner, subjective shift in motives appears now to be transcultural, but it does not necessarily lead to disengagement. The surgent potentials of later life lead to overt passivity and social withdrawal only when they fail to meet a sponsoring and facilitating psychosocial ecology; in the Druze case, Allah himself is the sponsor and facilitator, along with the community of prayerful men. His coreligionists recognize the worshipful potentials latent in the emergent passivity of the elder Druze; they move them toward sculpted expression in the *aqil*'s religious observance. In short, the Druze case shows that the surgent psychic developments of later life are indeed inexorable but not necessarily a prelude to social withdrawal and physical death; given a society that recognizes the emerging dispositions, values them, and gives them role articulation, the so-called passivity of later life can provide the ground for a later-life revival, a kind of social rebirth, even into leadership positions.

The Personal Redemption of the Traditional Elder

The Druze and other ethnographic data document the postparental advance of traditional older men at the second and third Linton levels, the social and the personal. At the personal level, we see them, via their enlistment in sacred systems, retaining and even gaining self-esteem, a renewed sense of self, despite the usual attritions, insults, and stigmata of aging. Thus, if traditional elders appear strange, eerie, it is because they are a bit like the alien gods, endowed with *tabu* power, not because they have taken on the stigmata of weakness and death. Moreover, by blending with the eternals, they overcome the awful sense of catastrophic, mortal change in their body and the sensory apparatus. In the traditional setting, the forces that waste and weaken men come from the gods. The old man is not castrated by his own failing body but burned dry, a nobler fate, by the exposure to divine power. His wrinkles are not the outward signs of shame but battle trophies, dueling scars from God.

The gods can be, to the older man, the ultimate male ally: the power that restores his flagging male potency and that buffers him against the assault from feminine identifications surgent in himself and from female power in the hands of his ascendant wife. In traditional worlds, in the community of folk, loss is not experienced passively but becomes the basis for a renewal of Active Mastery. As de Beauvoir (1972) put it, "the aged are raised above the human condition and have become immune to the dangers that threaten it."

The Old Man as Culture-Tender

We have also seen that the elder's special role—as power broker, splice between the folk and its gods—qualifies the old man in terms of Linton's second, or social, condition: The mana that the power-bringer freights in revives not only himself but also the natural and social organs of his community. He becomes, for his people, the prime source of luck, of success in battle, of ripe crops, of fat herds, of healthy children.

A viable society is founded on enduring structures—institutions that outlast the individuals who man their stations. As such, institutions have both formal and energetic aspects. They persist as recognizable forms by virtue of the neutralized energies that are continually directed toward their service. We have seen how old men serve as power transformers for their societies' structural grid. But in fully comprehending their social role, we must also take note of their crucial role, as culture-tenders, in creating and preserving the social forms, the institutions, that are maintained by these powers.

To better appreciate the special, culture-tending assignment of the postparental man, we must first acknowledge the eternal dialectic, the vital tension between cultural continuity and social change. Bear in mind that I am not a Pollyanna: The developmental view of aging does not rule out the dimension of loss. But many seeming losses of later life—much like those of earlier life—can be seen as the precondition for further advance. Men lose the physical capacity and psychological appetite for taking life, and women lose the capacity and appetite for giving life; but as the psychophysiological systems that serve the emergency conditions of adulthood—for example, war and childbearing—phase out, they reveal a previously hidden, vegetative physical and psychological organization. These changes uncover a body/mind format that is fitted to stable social and physical environments. Younger individuals have bodies and impulses that lead them to relish and provoke change; older people have bodies and appetites that lead them to relish and sponsor the equally vital dimensions of social stability and continuity. As Brian Griffin (1984) has shown in samples from the United States (and as our cross-cultural review confirms), the task of experimenting, of provoking social change and new social forms, is generally assigned to the young; the task of maintaining social continuity—the constant beat of the social heart—is generally assigned to the elders.

The candidate seeks the office as much as the office seeks the candidate. Old men are not only elected to their special positions; faced with often debilitating change in all life domains—physical, social, and existential—they are reactively driven to seek out and identify themselves with the eternal, unwavering center. Thus the statement of an aged Druze: "Everything in our life has changed; only Allah does not change."

The nature of this enduring center is conveyed by the anthropologist Robert Redfield's observation (1947) that a culture is a system of shared understandings: of what is good and bad, possible and impossible, thinkable and unthinkable. I would amend this: A culture is a system of idealized understandings. Ultimately, culture is what you will die for, or what you will willingly send your sons to die for.

It is this idealized and enduring aspect of culture that older men represent, making it real for themselves and their people. More specifically, older men are best equipped to identify themselves with the founding or origin myths of their culture.

Elders as Emeritus Parents: The Strong Face of Aging

This last assertion calls for a brief digression on the nature of cultural myths: Just as individual identity is based on a core sense of uniqueness, the sense of cultural identity and specialness is based on the legend of a unique history and of a special beginning. Culture is the unique meaning of an otherwise ordinary society.

All cultures conserve and amplify origin myths: stories that tell how the gods or unordinary individuals intervened in some special way at some crucial juncture, when the folk were in a vulnerable state, to bring about a special people, the favored children of special protectors. We think of Moses, the agent of Yahweh, bringing the children of Israel out from their oppressed state, leading them through the trial of Sinai to their national rebirth in Palestine. Or we think of George Washington, "father of his country," leading untrained militia through the trials of defeat and Valley Forge, until they achieved a victory against all odds to become Founding Fathers in their own right.

These sacred accounts, when enacted in ritual and daily behaviors, link the observing individual to that special time and to the special powers that moved through them. In its ideal or sacred sense, culture is a set of transformational rules that allow the average participant in the encultured society to locate himself within the myth and to understand his own conforming behavior as a working out of the mythic principles.

I contend that traditional older men, because of their special access to the gods, to the myth-embracing past, and to the spiritualized ancestors, become the living exemplars of the legend. It is they who bypass the time gap between then and now, bringing the two into conjunction. Via the elders, the mythic past and the mundane present are interpenetrated, and the climate of mythic origins is brought forward, into the here and now. Through the rituals managed by the older men, the power-shed of the past is tapped, and the mana that once rescued a desperate people reenters the world to heal the ill, to bring reviving rain down on a wasted land, or to move puny boys into strong manhood.

The traditional older man, by tending culture, serves society as much as he is served by it. He preserves culture by making its mythic armatures real and available for the larger group. Older men thereby help to provoke the experiences upon which social bonds are based: experiences of automatic familiarity, of common membership in a unique collectivity. For when separate individuals hold the same ideals, they become familiars to each other, automatic comrades, even

229

though they are not directly known to each other. Each participant in culture recognizes, feels *familiar* to other culture mates through self-examination, such that the other becomes a kind of self-extension and can be endowed with the narcissism, the self-love that would otherwise attach only to the isolate individual. By thus providing, for all age groups, ideal entities that facilitate the binding of narcissistic energies, older men facilitate the transformation of what is originally antisocial egotism into reliable social bonds, and what is potential anarchy into reliable conformity.

The traditional elder, then, helps to diminish the tension between the sacred and the profane; through him the sacred order acts to legitimize the rules of secular behavior. In the traditional culture, these are not merely arbitrary restrictions—"the games people play"—but the outer, active translation of the great traditions, the mythic core of the culture. In these settings, conformity does not imply, as it begins to with us, the shameful surrender of self-potentials; rather, the tie to the sacred order is confirmed through proper behavior. The deprivations and restrictions required by social living are thereby traded for meaning, even for heroic stature, and conforming behavior becomes the main route to self-esteem. Through the sacrifice of aberrant potentials, the proper individual acquires a sense of identity, of sameness, with the transpersonal, mythic order, and a quickening of self-esteem: the recognition of the legendary center within the self.

These special talents of aged men, to create and serve the cultural reality, are employed by traditional communities; but they also mature usefully in traditional subcultures and professions of modern, secular societies. The legal profession, particularly for those who seek public service, is one such traditionalist enclave; it is not surprising (though little noted) that the near-impeachment of Richard Nixon was accomplished by old lawyers, most of them in their seventies. Young men were spear-bearers on both sides, but the old lawyers—whether acting as judges, special prosecutors, congressmen, or senators—were primarily arrayed against Nixon. They opposed a powerful president who had dared to tamper with the Constitution, a document freighted with the mythic origins of this society. Old lawyers and legislators had taken the Constitution, an insubstantial tissue of words, and by an act of love and idealization made it real for themselves and for the nation as a whole. For many of them, the Constitution was a cultural reality in the fullest sense of that term. It was idealized and internalized; its adherents drew from it their self-esteem and a sufficient motive for self-sacrifice. In their minds, in their words and actions, the democratic principles that lie behind daily political life were manifested, made real for the nation. In effect, as in the traditionalist folk society, the old lawyers had moved to the sacred domain, the spiritual perimeter of the legal subculture; their audience did not see tired old men sitting in judgment on Richard Nixon, but the numinous presence of the law itself with a human face. Mustering the powers belonging to that great representation, they could bring a president down.

Culture-tending and the Parental Imperative

The social utility of the elder male, his service at Linton's second level, is now clear. But true development must encompass all the Linton levels. I will now review the ways in which postparental men and women serve, in their own way, the first, or species, level: the human parental imperative. Although I have concentrated in this chapter on the traditional older man, the roles of older women can help to clarify the masculine species-preserving functions.

I argued earlier that the unique vulnerabilities of the human child call for tremendous sacrifices in all spheres—physical, psychological, and social—on the part of the conjugal unit, the biological parents. It is true enough that, as a counterpart to their neediness, our children have tremendous potentials for new learning and high accomplishment, so that parents can in some part redeem their own sacrifices by dreaming of their children's future success. But fantasy is usually not enough. As we and our children discover, the unbuffered, unsupported nuclear family is too often not equal to the grinding burden of raising physically and emotionally viable children. Men have always been prone to defect from domesticity and to abandon children; but in our time unsupported mothers can also become child abusers. Thus, while the primary burden of child-rearing will always fall on the biological parents, they will not succeed in perpetuating their society or their species without various forms of outside support. In effect, mothers and fathers themselves need to be "parented," and each parent in his or her special way. Our species has evolved two great protections for the human parent: the extended family, which particularly supports the young mother, and the institution of culture, which particularly supports the young father.

As we have seen, the extended family, under the executive sway of the kin-tending matriarch, provides the young mother with a ready-made women's community: older matrons to guide, direct, and re-assure; sisters and sisters-in-law to share the physical and emotional work of the mother, and to provide built-in companionship against the isolation that child-rearing can entail.[7]

In this system, older women graduate from hands-on care within their own nuclear families to administration of the larger unit on which the decencies, the altruisms, of parenting depend. Clearly then, the older woman has not ceased all parenting; instead she has grad-uated from the emergency phase of the larger parental condition. As an emeritus parent, she now uses the aggressive powers that could have destroyed her nuclear family to safeguard the organizational integrity of the extended family. In short, as older women move from providing emotional security to guaranteeing social security, they re-claim their hidden aggression, deploying it openly in their senior pa-rental roles, their positions of eminence within the extended family.

By the same token, as older men become culture-tenders, they too support the necessary decencies of parenting and species renewal, in a special but most vital way. Older women, via their assertiveness, manage social life; they control the organizational contexts of par-enting. Older men, via the humility that ties them to the gods, con-serve culture, the idealized systems that relate the pragmatic social order to the legendary power-shed and the mythic ideals. In so doing, they provide the conformities and restrictions of daily social life with meaning and significance. When we recall that adequate parenting calls for great restraint on the part of women, and for the surrender of omnipotential claims in favor of the child on the part of men, then the relationships between culture-tending and parenthood, particu-larly fatherhood, become more clear. As Erik Erikson once remarked (in a lecture to Harvard undergraduates), deprivation per se is not pathogenic; it is only deprivation without meaning that is psycho-logically destructive. By providing idealized meanings, a strong culture gives significance to the imposed deprivations that are at the very heart of parenthood.

Domesticity and parenthood carry meanings that make these states particularly onerous to men; in the most general sense, cultures function so as to render the life of the householder palatable to men. Women, as young mothers, have less trouble than men in entering the domestic world, for this is a domain that, in traditional societies, they have never left. On the other hand, the early socialization and testing rituals experienced by men split them away from the domestic world, the terrain of the mother, and point them outward, toward the perimeter and the father's range. Reversing their social gravity, their own fatherhood pushes them back toward the world of women. Grown men can better tolerate the encounter with domesticity if their conforming actions are given meaning: preferably a heroic significance, to compensate for the loss of the free bachelor life, on some version of the physical or social frontier. For example, American fathers were more reliable parents during the Great Depression than during the periods of affluence that followed. During hard times any father who kept his family fed and housed could, with reason, feel heroic and irreplaceable. Paradoxically, hardship ennobled the father's role, and provided him with compelling reasons to stay in it.

But such meanings are usually provided by ritualized hardships, rites of passage instead of social disasters. In early life, in the puberty rituals managed by their grandfathers, boys endure ordeals to gain strength. Spent and weakened, like their own people in the origin myth, they are rescued by some totemic sponsor. By living out their people's legendary cycle of despair and salvation, they enter into the texture of the myth, thereby gaining the sense of resource that they needed to separate from the mother. Then, proving themselves as men among men, in terms set by cultural values and ultimately by the founding myths, they gain the sense of masculine identity that allows them to reenter the domestic world without feeling castrated or infantilized.

In short, it is the man's life in culture, his role as it reflects the founding myth, that allows him to tolerate both the physical risks of the perimeter and the emotional dangers of the reentered domestic world. The ritual life that old men, as culture-tenders manage, gives young men the courage to leave the mother and to return, as fathers, to the mother (as personified in their wives and, covertly, in themselves). Old men help to provide young men with the powerful meanings that they require in exchange for giving up the temptations of barbarism and random procreation in favor of civility and fatherhood. When old men, as keepers of the flame, tend culture, they, like old women, are still engaged in parenting: In them, the late-life turn toward androgyny unmasks the latent skills and traits that fit them for special, gender-specific roles in the service of cultural stability and, by extension, adequate parenting.

The two great support systems of human parenthood—the extended family and the traditional center of culture—are as a general rule headed by strong elders, men and women who have developed, *de novo* in their later years, the capacities required for these great assignments. It is these genotypic, specieswide maturations, rather than fortuitous adaptations to loss, that constitute the developmental phenomena of later life.

Finally, note well that postparental growth potentials become more evident when we take a transsocial and even a transspecies sighting on later life, expanding our samples to include postreproductive primates, as well as postparental men and women. As one consequence of this review, we now realize that the very term "postparental" is inexact; given properly facilitating circumstances, older individuals become emeritus parents rather than former parents. They do not lose touch with the suprapersonal, species aspects of the life cycle; instead, the surgent potentials of later life bring about new connections with species goals, moving older people toward new assignments, wherein they maintain the larger social and cultural frameworks of effective parenting.

In the next and final chapter, I will look at the consequences—for emeritus parents, for biological parents, and for children—when culture and the extended family are diminished, when elders lose the usually traditional settings in which their special form of development can go forward.

10

DECULTURATION AND THE PASSING OF THE ELDERS

An aged man is but a paltry thing,
A tattered coat upon a stick, unless
Soul clap its hands and sing, and louder sing
For every tatter in its mortal dress,
Nor is there singing school but studying
Monuments of its own magnificence;
And therefore I have sailed the seas and come
To the holy city of Byzantium.

O sages standing in God's holy fire
As in the gold mosaic of a wall.
Come from the holy fire, perne in a gyre,
And be the singing-masters of my soul.
Consume my heart away; sick with desire
And fastened to a dying animal
It knows not what it is; and gather me
Into the artifice of eternity.
                    —WILLIAM BUTLER YEATS, 1865–1939
                    (from *Sailing to Byzantium*)

In the previous chapter I have argued for a conclusion that is quite startling to the modern sensibility, which accepts, as a given, the weakness and victimization of the aged. I have argued that, rather than being victims, the aged—particularly aged men—are the authentic heroes of the folk-traditional society; and that their special stature has to do with their privileged access to supernatural power sources that they, in effect, create and make real for the rest of the society. And I have argued that, in their kin- and culture-tending roles, older women ease the parenting trials of young mothers, while

235

older men, by imbuing conforming behavior with mythic, heroic meaning, make domesticity palatable for young fathers. In this chapter, again evoking Linton's levels, I will sketch out the effects of rapid social change on the personhood, the social stature, and the species assignments of the elders, with particular reference to the aged of our own society.

Modernization, Urbanization, and the Erosion of Gerontocracy

I argued earlier that the "natural" form of human governance, repetitive across the most enduring forms of human settlement, is gerontocracy: matriarchy for the extended family, elder patriarchy for the more public and ritual side of traditional folk life. Survey research and ethnographic vignettes from around the planet give clear evidence of the erosive effects, particularly on elder male leadership, of rapid modernization, industrialization, and urbanization.

Politically speaking, modernizing and urban societies (the two are not the same) do not abandon male gerontocracy per se; the majority of their rulers are, as in the folk-traditional situation, older males. The shift with modernization is away from participatory gerontocracy, the condition in which honor and power are afforded older men merely by virtue of their being old. By contrast, the elder rulers of "advanced" societies are only older versions of successful young men, those who have laid down the bases for their economic and political power early in life. They receive little honor, title, or credit on the basis of their age alone. For the rest, for the majority of undistinguished older males who have not laid up power and riches in their early years, personal and social prospects can be bleak, making for a striking and disheartening contrast with the typical older man of the folk-traditional assemblage.

The Economics of Modernization

The losses in senior male supremacy along the folk-modern continuum are evident; less clear are the reasons for this great slippage. As discussed in the previous chapter, the power of the old man in the traditional community rests on a number of bases: social, develop-

236

mental, and existential. The social guarantors of the aging man are his accumulated wealth, his membership in an extended family, and particularly his identity with the cultural mystery. Developmental factors have to do with the evolution of special mental capacities (wisdom) and the conversion of surgent potentials into ego capacities that serve personal satisfaction, social continuity, and species survival. Existential factors have to do with the qualities of "strangeness" and physical debility that are translated, by the sacred organs of culture, into the currency of *tabu* power. I will deal with each of these factors as they are modified by two different social conditions, modernization and urbanization.[1]

Turning first to the effects of modernization on the economic power of the aged, we find that the modernizing society typically provides a labor pool for the developed world; the young men, instead of staying home to work their father's fields, leave the traditional village to find wage employment in the plantations, mines, mills, and armies of the more developed sectors.

As a consequence, the young men discover an economic base of power to which they have privileged access and which is not under the control of their fathers. They no longer have to look to him for the bride price, for the economic substance that will accelerate their change from boys into men. The confirmatory power now comes from outside the precincts of the community, rather than from within it. As a cogent example, Rowe (1961) finds that village gerontocracy in India has eroded as a result of modernization. The young men find wage work in the cities and return home impatient with tradition, with the rule of the father, and with the slow-paced village life. Furthermore, the wealth that they bring back undercuts the traditional association between affluence and advanced age, and further corrodes the status of the aged. Irwin Press and Mike McKool (1972) find a similar effect in the villages of Mayan Mexico, when the *avanzado* (progressive) elements of the traditional village draw on outside funding to build the small village factories that will produce new forms of capital. Shifting from agricultural to wage work, the young men can now buy their own land before their father is ready to surrender the family holdings to them. In his dotage, the father now comes to live in the son's house, rather than granting the son's family a place in the patriarchal compound. As a result, the once mighty leading men, the *principales*, are degraded, and aging fathers who hold onto their land, refusing to recognize this shift of power, are sometimes murdered by their own sons.[2]

Modernization and the Loss of Wisdom

Stable, insular, rural societies are hothouses for the flowering of senior wisdom. There, the cultural assumptions go unchallenged, predicting in a rough way to actual events; and so the old men, expositors of the cultural themes and principles, develop a reputation for omniscience. Not surprisingly, the loss of rural isolation commonly causes the first shock to the unquestioned rule of the "wise" gerontocrat. So long as traditional beliefs and values can flourish in their own niche, uncontested by alternate world views (particularly those emanating from more developed societies), they avoid the fate of being relativized and called into question. And so long as the cultural ideas remain inviolate, their elderly representatives keep some commanding stature, as wise men. But when a society moves out of its isolation—or is moved out of it, by conquest or colonization— the people are automatically brought into contact with contrasting and even conflicting conceptions as to what is good and bad, possible and impossible, holy and impure. Despite this disorienting and relativizing contact, the emerging society may retain much of the ideological content of its culture; but the structural relationship between the people and their special beliefs must change as a result.

The contact with foreign conceptions eventually brings home to the once-isolated people the realization that their particular system of cultural ideas is only one among many such systems, some of which may be more powerful or more proximate to the truth than their own. Their unchallenged beliefs become relativized; worse yet, they become revealed as nothing more than ideas; they lose their specialness as the unquestioned coordinates of existence. Under such conditions, the cultural conceptions may still be widely distributed, familiar to most affiliates of the society; but now they are known as symbols, not as substance. They are no longer the basic, power-compelling axioms of the universe. And as the ideas become relativized, as symbol is split from substance, the old men who represent and expound the ideas are likewise split from their special powers.

The shock is most severe when the emerging society makes contact with the beliefs and practices of a technologically advanced society, for their engineering and medical skills bring about immediate, tangible "miracles" that cannot be matched by prayer and ritual. Thus the power of the Navajo medicine man has been broken not by white missionaries but by Anglo medical interns in the Public Health Service hospitals scattered across the reservation. Despite all his awesome dignity, despite being decorated with silver and turquoise, the medicine man's prayers cannot save a child with intestinal infections. The singer will chant powerfully for nine nights, while the child dehydrates and dies. Responsive to the parental emergency, the Navajo have learned that a callow white intern, who commands penicillin, can save the child that the dignified medicine man will lose. Accordingly, at-risk children survive, but the mythic basis of Navajo culture is at the same time called into question, and Navajo lifeways lose their power to provide individual Navajo with sacred sentiments, the sense of significance and self-esteem.[3]

Modernization and Mythic Power Sources

While the majority of anthropologists, in their practical-minded fashion, concentrate on the generational shifts in economic power that are typically brought about by modernization, the loss of wealth is not the independent variable in the patriarch's decline, just as it is not the independent variable in his ascendancy. "Wealth" in the folk mind is important only insofar as it links up with traditional ideas about *tabu* power. Wealth in material goods is important only insofar as it represents favor from the gods (or the power of witchcraft used in the service of acquisition). Material wealth is the by-product of ritual power, and such *tabu* power tends to be the major goal of acquisition in later life. It is not the generational shifts in wealth and ownership per se that affect the status of the aged; rather, the reduction in senior wealth and property brought about by modernization has crucial consequences for the generational distribution of sacred power, for the access to such power on the part of older men, and for the very conception of power itself.

With modernization, the traditional culture, which stresses the ritual access to sacred power, loses its commanding position within a special preserve, even as the new forms of power made available and significant by modernization take on some of the vibrancy that the traditional forms have lost. Power becomes secularized: represented by cash, by machines, by the products of machines, and even by beverage alcohol. In effect, with modernization, the distinction between totemic and secular power phases out, and the power boundary is reset, no longer dividing the sacred from the ordinary world but instead standing between the "underdeveloped" community and the "modern," or Western, world. The aged are the monitors and heroes of the old power frontier that faced gods and demons; it is young men who cross the new frontier to bring back alien power, in the form of wealth and technology, for themselves and their community. This is a role for the promethean young and not for the cautious elders. If anything, the role of the old men on the new frontier is a reactionary and even destructive one. They tend to distrust the new, deritualized sources of power; because of their traditionalism, they impede development and the flow of significant power from its "modern" reservoirs into the still relatively powerless, emerging community.

These changes, of course, occur slowly. The old man may be feared as a sorcerer long after he ceases to be respected as a healer; witchcraft beliefs are always the last to go. But as regards the self-esteem of the aged, the important change has taken place: The older individual is no longer a benefactor, a source of good rather than bad energy.

As the aged lose their cosmic connection, they also lose the power to control the cultural imagery in their favor. New images and new myths, based on the power of aliens and of contemporary heroes who stole alien power, fill the cultural space. The young can interpret and fit into these myths, for they have been in the place of the alien: They have endured his power, and they have—like Prometheus—captured some small part of it in the form of wages, technical knowledge, and even weapons. The young begin to take on some of the glamour of those who have danced before the gods and have survived the awesome contact with strange power. As we will see in the examples that follow, when the young take on the bridgehead functions vis-à-vis alien power that the aged once performed, the prestige of the young is elevated, and the prestige of the traditional aged correspondingly declines.

The demythification of the elderly takes place in tandem with the demythification of culture as a whole. For example, in our own country, Andrew Achenbaum (1974) finds that the historic movement toward American modernity had the same consequences for the aging that are brought about by modernity generally. Prior to the Civil War, in the period 1790–1850, the aged are presented in public prints as venerable, and as socially, morally, and economically useful. Furthermore, they were repositories of wisdom, guardians of the revolutionary spirit. They were commonly portrayed as having one foot in heaven

and sends some of them home to challenge the local gerontocrats. By the same token, particularly during its period of rapid gestation, the city itself marks the later stage in a revolution against the rule of the rural gerontocrats.

Thus the massive youthful migration from rural hinterlands to cities is one of the most pronounced and predictable cross-cultural phenomena. During their growth spurt, cities are magnets for young men who go there in search not only of wage work but also of adventure, and of an exotic milieu where they can experiment with strange women and forbidden lifeways, without having to make final choices among them. But finally they go to the city as a kind of protest against gerontocracy, against oppressive patriarchy, and against the omnipresent extended family—the ring of monitoring, critical, gossiping elders. Accordingly, the psychic dislocation of the elders (and particularly older men) in the city may represent more than an accidental side effect of urban social organization. It may be by design that the structures of city life function to prevent the recurrence of male gerontocracy, and to give fuller expression and social power to the oppressed constituencies of the patriarchal village: women and young people of both sexes.

But liberation is a two-edged weapon, and the forces that are set free in the city can act so as to diminish men. Besides giving power to oppressed social constituencies, the contra-gerontic genius of the city also permits the liberation of oppressed psychic energies, the range of erotic appetites, mature and primitive, that are kept in check under the rule of village elders. We have already seen that the index of syntonic orality rises in step, for Navajo and Druze men of all ages, with the proximity to cities. But the syntonic orality index measures not only vital appetites; it is also a valid register of male passive-dependence. As previously noted, it correlates with passive themes in dreams and the TAT. More to the point, orality is related to alcoholism, to the severity of physical illness, and to emotional vulnerability. In short, the oral index of male passive-dependency—hence, of psychic vulnerability—appears to rise in step with the residential proximity to cities. Older urban men can become victims not only of their own oral need but also of the selfish, omnivorous urban young.[4]

Urbanization and Matriarchy

Our tentative conclusions concerning the psychological depletion of men in urban settings are strengthened by the findings in regard to women. While urbanization appears to sponsor male passivity, it has the opposite effect in regard to women and spurs their liberation. The modified-extended family of the growing city provides less by way of security and guarantees against estrangement for men than it does for women. Thus Elmer Youmans (1967) concludes that men are likely to be the recipients of emotional comforts in rural areas and that this pattern is reversed in the cities. Likewise, Ernest Burgess (1960) notes that the urban family is based on strong bonds between grandmothers, mothers, and daughters; and that the matrilocal and modified-extended family can function in stable working-class neighborhoods as the urban equivalent of the patriarchal village of the countryside. In the city, Burgess says, "men have friends, women have relatives." This observation is echoed by Michael Young and Mildred Geertz (1961) who asserted, following a review of family ties in San Francisco and London, that "the mother-daughter bond is the central nexus of kin ties in industrial societies."[5] Not surprisingly, in the urban American setting, men die younger than women, and, as we learned earlier, men are more likely than women, at any age, to commit suicide.

Thus, while superior female longevity is sometimes taken as a sign of women's innate biological superiority over the male, this difference may actually reflect nurture rather than nature: It may register the peculiarly troubling effects that the urban milieu seems to have on the psychological makeup of men.

In sum, with the growth of cities and the movement of peasant populations into cities, there is a loss of male patriarchal prestige, an increase in male somatic vulnerability, and an increase in male oral appetites, as well as in the more behavioral expressions of male passivity. At the same time, there is a clear increase in matriarchy, female longevity, and female liberation. In effect, rapid urbanization replicates the psychological effects of aging: It releases the male tendency toward passivity, and the female tendency toward dominance, outside of the family as well as within it. In the city, men and women conserve narcissism for themselves and strive after omnipotentiality. Thus young urban men preserve their oral and passive urges, and young urban women preserve their capacity for aggression.[6]

Urbanism: The Aging Man as Stranger

Besides revealing an "unbuffered" male tendency toward passivity, the city environment also unmasks other noxious consequences of the rural-urban shift, even as it reduces the protections against them. The urban milieu has the effect of revealing and underlining a preexisting stigma that always attends the aging, the mark of the stranger. Earlier, I cited de Beauvoir's observation that the aged are always in danger of becoming the stranger, the "other"; I traced the ways whereby the sacred aspect of traditional culture functions to transform "otherness" into a version of elderly grace. Three central characteristics of urban life work to bring about this degradation, from the status of revered elder into one of the aged as stranger: the weakening of popular or consensus culture in favor of the high culture of elites; the weakening of the extended family in favor of the nuclear family; and the substitution of an idealized collectivity by an idealized self: the urban sponsorship of unequivocal narcissism. I will discuss each of these, briefly.

Culture and the Protection
Against Alienation

In the traditional society, the aged are protected against the stigmatized meanings of their strangeness not only by particularly favorable cultural conceptions but also by the institution of culture itself. The traditional society sponsors the generation and propagation of gerontophilic ideas; but it also sponsors the vitality of culture per se, as a set of idealized regulations that, because they are internalized by individuals, ensure predictable group behavior in conformity to the cultural prescriptions. As a consequence, within the high-consensus traditional society, familiarity tends to be automatic and does not require direct association. Through knowing yourself you can know, can predict, the other who shares your culture, even though he is not your acquaintance or your neighbor. The aura of "we-ness" is extended to those individuals, of any age, who speak, dress, and act predictably, according to the group consensus. The true stranger is found outside, not within the community. The cultural system of the traditional society, irrespective of its particular beliefs, plays the great role of making even potential strangers responsible for each other and of converting narcissism into social bonds. The resulting endowment of protection extends to the aged: So long as they are included under the umbrella of "we-ness," they can preserve their humanity in the eyes of younger people, and in their own eyes, as well. But the urban environment does not as a rule collectivize and standardize personal experience so as to support the basic postulate of culture: that the other can be known and predicted, through reference to either shared custom or the self. Whereas the society of the traditional village society creates familiars, the city pools strangers, those who remain unfamiliar and unpredictable despite years of physical proximity.

In effect then, modernization erodes the cultural ideas that favor the aged, and rapid urbanization, which undercuts the standardizing and familiarizing power of culture per se, may destroy the aging person's greatest defense against his possible estrangement. In the countryside, the aged are made familiar, and their quotient of strangeness is converted, via the sacred system, into awe and reverence; but in the city, where weakened culture loses its power to ensure familiarity, all urbanites move toward the condition of estrangement. There, the "stranger" potential of the aged is maximized, in conditions under which strangeness cannot imply a connection with sacred power. Thus, whereas awe of the aged, *because* of their strangeness, predominates in the traditional countryside, revulsion against aging and the aging stranger tends to predominate in our cities.

When urban culture loses its power to bind and transform narcissism in all age groups to the collective weal, egocentricity increasingly becomes the general coinage of social relations. The aged, who depend on the goodwill and on the inner controls of younger, stronger individuals, as well as on the external social controls that are maintained by a strong cultural consensus, suffer accordingly. Urban deculturation not only underscores the negative meanings of their strangeness; it also brings about an increase in the numbers of urban barbarians who victimize the aging because of this same estrangement.

Deculturation and the Generational Schism

Furthermore, in our country, and particularly in our urban settings, we threaten the integrity of culture in particularly American ways, practices that have uniquely drastic consequences for our civil life, and especially for our aged, within their families. The egalitarian ideal of our culture eventually brings about consequences that can even destroy the founding culture. Thus the impulse toward equality, which first struggles against the restrictions on ordinary civil rights, ends by politicizing the complaint against existential restrictions. Individual Americans were long ago granted the right to choose their leaders, careers, and political parties; but now the principle of equality attacks our cultural assumption, of common unquestioned values. Until recently, our cultural values were regarded as objective: fixed social realities to be internalized by the individual, but to be created and maintained apart from the same individual, by idealized cultural institutions and their leaders. In recent times, the process of value formation has been democratized, taken out of the institutional province and given over to the individual. Moving thus, from the social to the personal sphere, values lose their shared and objective character, to become private and subjective. In effect, we have democratized and relativized the process of value formation to the point where each citizen is conceded the right to decide personally what values should be and by what standards the individual should be judged. Outside of a courtroom, the assertion, "I'm just doing my own thing," comes to be the ultimate justification for any kind of behavior. In the obsessive pursuit of equality, we have succeeded in subjectivizing values to the point where weakened culture no longer coheres shared ideals. Culture loses the power to convert strangers into familiars, and narcissistic potentials into enduring social bonds.

When culture is strong, individuals relate to each other in terms of the explicitly social categories that emerge readily out of group experience and affiliation: family, neighborhood, ethnic group, religion, social class, profession, and nation. The groupings formed under these categories admit individuals to membership regardless of their age and, to a large degree, regardless of their sex. When culture is strong, the social categories, formed out of collective experience, predominate over those formed out of individual, idiosyncratic experience, and determine the lines of affiliation. When culture is strong, those of my social class are my automatic allies; all Americans are my fellow countrymen. But under the conditions of deculturation, narcissistic preference dictates the lines of affiliation. I can extend the awareness of familiarity and selfness only to those who are like me in the most concrete, asocial, immediately sensible respects: those who share the same skin color, the same body conformation, the same genitalia, the same sexual appetites, and the same age group as myself.

In effect, with deculturation, the principles of association are no longer based on shared standards but become instead racist, ageist, sexist, and homoerotic.

Exclusive age, racial, and sexual groupings, based on narcissistic choice, are becoming the norm in our society, particularly in our urban centers. The consequences for American family life are especially drastic. While the family once united its members, regardless of their age position, the American family is often riven along fracture lines of generational and even sexual differences. To the degree that family solidarity still persists across the generations, it is mainly between the grandmothers, daughters, and granddaughters of the female line, more and more excluding of men. And by the same token, the mistrust between generations that is now a feature of our general social life has invaded the heart of the family itself. Typically, the family, whether nuclear or extended, has been the one great enclave in which generational distinctions were usually overlooked in favor of some higher principle of kinship solidarity. But the "generation gap" that is strikingly absent in the traditional family bisects the heart of the nuclear and modified extended family in the city: The aged begin to know alienation and even "strangeness" within their own families.

Because they are no longer legitimized by some sacred principle, cultural rules themselves come to be regarded as restrictions, as arbitrary games, rather than as routes to power, personal significance, and self-esteem. Soon enough it becomes apparent, particularly to the young, that honor and conformity are incompatible, and that power is to be gained through antisocial rather than prosocial acts. As the rebel and the psychopath acquire glamour, the social order, which depends on automatic trust among strangers, is increasingly compromised; and the urban aged, who depend on a stable social order, are increasingly put at risk.

De Beauvoir (1972) has observed: "A surer protection—than magic or religion—is that which their children's love provides their parents." The narcissistic ethos of the American city tends to weaken that great protection and to increase the vulnerability of the aged, thereby making loss and emotional deprivation the great theme of mental illness in later life. In the city, the defection from the aged by their children has reached the point at which the provision of extrafamilial and institutional forms of care to the aged has become a major social task and problem. More and more, society is called upon to provide the physical and even the emotional security that urban children are less willing to provide their own parents.

One major result, afflicting both sides of the generation gap, is a kind of pseudohedonism, in which young and old look to impersonal substances and gratuities—drugs, video, liquor, food, cults, therapists, charismatic leaders, casual sex—to provide the feelings of comfort and security that they can no longer trust kinsmen or intimate friends to provide. Under conditions of deculturation, addiction becomes another substitute for true intimacy, a substitute for the self-esteem and the sense of inner substance that are no longer provided by the extended family or by culture.

Urbanization: The End of Maturation

While the modified extended family may partially substitute for culture in preserving the humanity and, by extension, the mental health of older urban women, it does not play the same role for older urban men. They tend to lose the protection of culture, particularly after their retirement from the work society, and to get less protection from the extended family.

Rapid urbanization, particularly when compounded with modernization, does leave intact some components of the older woman's milieu, such as the modified extended family, while virtually destroying the myth-imbued, traditional sector: the maturational setting for the older man. The extended family, although it undergoes modifications, is one of the few human institutions that survives the transition from the countryside to the city; and so older women, particularly those from organized ethnic groups, still show some movement toward confident matriarchy. In their case, established extended families provide a ground in which the surgent aggressive potentials are recognized, sponsored, and converted into their sculpted form, as the energy and drive of the kin-tending matriarch. But in the case of older men, the high culture that flourishes in the city is directly antagonistic to the traditional culture. Indeed, the high culture of creative literature, experimental theater, academic criticism, etc., gains its morale and vitality by destroying, symbolically murdering, the traditional myth-centered culture, often replacing it with an ahistoric secularism. The high culture makes its agenda out of questioning certainties, of exploding hallowed myths, of ridiculing conformity, and exposing the frailties of commonly idealized figures. The high culture kills off

249

the traditional culture for the same reasons that young men attack gerontocracy: in order to remove the "dead hand" of the past, the established precedents that block curiosity, creativity, and the experimentation with new, undreamed of pleasures. But the "oedipal" attack on tradition may achieve its ultimate goal in the debasing of the older man, who loses, in the destruction of the traditional culture, his special domain for personal growth, social prestige, and species utility.

By contrast to the older woman, the aging urban man—unless he is fortunate enough to be enlisted in a traditional subculture (or profession)—loses touch with the sponsors and facilitators who could help him redirect his surgent dependency in legitimate and dignified ways: toward the service of the empowering gods, the institutions that guard their ways, and the rituals that tell their story. In the absence of such sponsorship, the older man's problematic strivings toward dependency are less likely to achieve a developed, sublimated form and are more likely—as is the case in our society—to fuel the psychological vulnerabilities of later life.

The Casualties of Elderly Narcissism

The failures of later-life development leave older men vulnerable not only to the narcissism of others but also to their own. Aged Druze, who have converted narcissism, the idealization of self, into worship of their god, can look with equanimity and dignity on the prospect of their death, saying only, "This is Allah's will; to complain about my sickness or death would be to dispute his will." The aged Druze have found a posture, based on the idealization of and identification with their changeless god, that permits them to transcend, without much sense of shock or insult, their awareness of inevitable change and oncoming death. But what of the decultured elder, who remains in the egocentric, self-focused position, without the possibility of transforming idealizations and without stable anchors against the winds of change, particularly when he or she is in the period of maximum somatic, cosmetic, social, and existential losses? If the focus on self is maintained in late life, the result is an increasing sense of vulnerability that transforms the "normal" losses and changes of aging into insults, outrages, and terrors. Depression and hypochondriasis, as well as strained and even delusional attempts to deny depletions and imperfections, can be the result. In our clinical researches at Northwestern University Medical School, we note that many emotional disorders of later life represent attempts to overcome the sense of insult and depression on the part of older individuals who have not relinquished the conviction of their own centrality and omnipotentiality. When psychiatric breakdown has occurred, the need to deny loss and insult, or to project onto others the responsibility for imperfection, is so strong that reality is abandoned in favor of the defense. Thus denial of loss and threat, sometimes taking the form of manic psychosis, is one major characteristic of the older patient. By the same token, paranoid states in which the sense of blemish, or the responsibility for the blemish, is projected onto others, is another strong possibility. Severe depression and lethal illness can result when these primitive defenses fail. In a real sense, psychosis in later life represents a hectic attempt to conjure up the sources of self-esteem that are routinely supplied to traditional elders by a strong culture.

In sum, deculturation and the weakening of the extended family have major effects that particularly disadvantage the aged. They lose their special developmental milieus, they lose their special bases for self-esteem, and—as a secularized culture loses its power to ensure civil behavior and to control the grosser manifestations of narcissism and rage—the aged lose their traditional character as hero, and take on their modern character, as victim. The weak face of aging appears.

Elder Abuse and Child Abuse

The aged are particularly hurt by the social entropy that results from the continuing revolution against the familial and cultural sources of gerontocracy, a revolution that begins in the modernizing traditional village but that reaches its culmination in westernized cities. But there is a paradox: The revolution against gerontocracy, intended to liberate the young, ends by putting children at risk. The two great social structures, extended family and culture, that protect and generate strong elders, are precisely those that protect human parenting and underwrite healthy children. By providing meanings that compensate for deprivation, culture helps make the human family and adequate human parenting possible. Given a strong culture, young adults will routinely, even cheerfully, enter into the chronic emergency of parenthood, giving up their narcissistic claims for omnipotentiality in

251

favor of their children's future growth. But as culture loses the capacity to endow deprivation with significance, children, as well as the vulnerable aged, are among the first to suffer from the primary narcissism and aggression, released by deculturation, that the elders once helped to hold in check. When culture and the extended family are no longer tended by strong elders, the unsupported, isolate nuclear family no longer remains a staging ground for child development but instead too often becomes the setting for child abuse, physical as well as emotional.[7]

Soon enough, as secularism drifts toward urban anarchy, the gerontic tragedy becomes the common tragedy. When the elders are diminished, much of our common life is eventually diminished with them. And as undisguised narcissism tends increasingly to become the coinage of urban social relations, all unproductive, dependent cohorts—children, the handicapped, the aged—are put at risk. Ageism and child abuse both increase in the city, perhaps at the same rate. When the motif of human relations becomes "me (or those like me) first," then the weak and needy of any age tend to go under. Geronticide and infanticide, elder abuse and child abuse, predict to each other across societies. The final outcome of rapid modernization is that the aged, once the automatic, entitled rulers of society, come to suffer the same vulnerabilities as children. A narcissistic age, that turns against procreation and parenthood on the grounds that children interfere with adult pleasures and that they are preproductive, also turns against the aged, on the equivalent ground that they are postproductive.

The fatal result is that many aged die of a kind of benign neglect in nursing homes, while we become uneasily aware that our cities have become the setting for a largely unreported holocaust of children: as a result of suicide and gang wars, but also at the hands of pimps, pushers, perverts, and serial murderers. Ultimately, children are the major victims of the indulgent, revisionist, relativistic games that we play with our culture. Again, while the crisis of meaning and civility that comes with deculturation has special consequence for the aged and perhaps touches them first, it is also very much a shared affliction. In our haste to be modern, in our revolutionary rage against tradition and gerontocracy, we have brought down the fathers and humbled the aged. In so doing, we are also bringing down the culture that sustains us all.

Redeeming the Elders

This is not to argue that we can benefit the vulnerable members of society only by returning to traditional folkways and patriarchal culture. It is far too late for that; besides, we know that traditional gerontocracy entails the disenfranchisement of women and the young. But as an unintended consequence of our American penchant for the full expression of the egalitarian principle, we may possibly end by filling the gap left by culture, for all its oppressions and inequalities, with the iron directives of totalitarian leaders who will be far more oppressive than the gerontocrats we left behind in the various old countries of our origin. The gurus, the Mansons, the Jim Joneses, the televangelists are waiting to "meet our needs." If we are to avoid that consequence, we have to find ways of reviving elderhood without reviving repressive gerontocracy. We have to at least begin by recognizing that our social malaise has its basis not in the change of culture but in the loss of culture. We will have to enlist the elders, who have traditionally been the wardens of culture, to help and guide us in the vital processes of reversing deculturation and of crafting the new myths on which reculturation can be based. We owe this redemption not only to our aging parents. We also owe it to the oncoming generations of children.

APPENDIX

Culture and Personality Among The Navajo, Druze, and Mayan Peoples

Navajo Personality: Surgent Features

Major themes in Navajo personality contribute to the prestige of the traditional Navajo elders. Regarding psychosexual matters, we can say that Navajo personality is organized around a keynote theme of oral aggression; there is much reference in their interviews to food, drinking is rampant, and envy of others' wealth or luck is a major theme, as is the fear of envy emanating from others. The folk-traditional society (especially when association is based on clan and extended family ties) tends to diminish ego boundaries, the sense of separation between the "I" and the "you," and these self-other distinctions are also undercut by the oral-aggressive position of Navajo personality. Both influences lead to a "zero-sum" view of the world, to the oral calculus that states: "If he has that which I want and lack, then it was taken from me, and I am justified in taking it back; and if I have what he wants and lacks, then it was taken by me from him, and he is justified in a vengeful attack against me."

As a consequence, the Navajo are quick to assume that any misfortune or illness is brought on by the witchcraft practices of a powerful person who has become envious of their wealth in sheep, horses, and "hard goods" (turquoise and silver jewelry). Thus the Navajo world is personalized, often fraught with a sense of sinister presences: Night noises on the roof could be the footstep of the *yenalchlonje*, the "skin-walker," the powerful sorceror who puts on a wolf's skin to speed across the darkened land. The need for protection against a world charged with such dangerous forces pushes the Navajo toward reliance on the interceding, generally older medicine men.

Given the porous quality of Navajo ego boundaries, other persons, external by our standards, become portraits of the rage, fear, or love that they can excite. In the Navajo world, objects do not have fixed, standard qualities independent of the personal emotions and idiosyncratic meanings that they call up. Rather, objects are defined outward, egocentrically from the self; they exist in the mind and eye of the observer insofar as he has feelings and motives toward them. For example, the Navajo fear the spirits of the dead to an extreme degree. According to traditional Navajo thought, death results from witchcraft or sorcery rather than from natural causes; the dead are so hedged about with taboos that the survivors will abandon the *ch'indi hogahn*, the house in which a relative has died. The dead persist as ghosts because strong feelings toward important persons do not end with their death; rather, they *resonate*, and so the Navajo dead continue to exist psychologically in the mind of the bereaved and socially in the structure of the community that knew them. Incorporeal as the undiminished feelings that insist on and sustain their reality, the dead take on the status of "ghost." Since traditional-minded Navajo do not see much difference between the psychic interior of the person and exterior reality, dreams of the dead are not regarded as symbolic events. Dreaming of a dead person, it is as if they find him concretely within the precincts of their own head, and they assume that the spirit has been introduced there in some forceful fashion.

For example, a venerable Navajo medicine man, a specialist in the ceremonial against ghost-sickness, explains this infection as the intrusive act of the ghost "wind": "The soul of the person resides in his wind, and this goes out of him when he dies to become the *ch'indi*. The *ch'indi* wind gets on them. Maybe sometimes it brushes them, but that makes them sick." In effect, the Navajo seem to experience the split-off, destructive aspects of their feelings toward the dead in the person of the malignant, vengeful ghost. In the traditional treatment the patient is purged so as to vomit out the ghost and is coated with tar so that the ghost cannot reenter. Finally, the ghost's vengeful spirit is bought off with the "ghost's bribe" (*ch'indi be'el*).

Navajo Personality: Ego Aspects

Thus far, I have described the contents of the Navajo unconscious and ways in which these are refracted in the projective systems—myths, healing practices, and witchcraft beliefs—maintained by Navajo society. While the Navajo may fear sorcery from faceless strangers, their behavior in more intimate and familiar settings reveals a different cast of personality, one sculpted by cultural norms rather than by unconscious fears and appetites. In these settings, individual Navajo tend to be playful, humorous people who take pleasure in nuances of language, in punning, and in provocative and even lyrical descriptions of action, movements, and places. The Navajo, once part of the Apache cavalry, delight in movement and in the language of movement; as they describe travels through distant places, their hands swoop out, to carve horizons. Even though they no longer ride horses on the raid, and although the pickup truck has replaced the pony as transportation (even in sheepherding), the horse is still the true totem animal of the Navajo: next to the Nine Nights sings, the Western rodeo is their major ceremonial. I have interviewed ninety-year-old men who claim that they had ridden a horse until age eighty-five, "until I got thrown."

As an oral people, the Navajo also "eat" the world through their eyes; their interest in motility has to do with the fact that movement opens the world and creates a large smorgasbord of visual pleasures. Love of the beautiful permeates many Navajo activities and materials and finds its boldest expression in their striking jewelry, textiles, sand paintings, and the verses of their healing chants ("sings").

The Navajo are conflicted between their envious cupidity and their love of beauty and expansive movement. As their religious beliefs suggest, they live always on the brink of disharmony—not only between themselves and the world, but within themselves. The old men play an important role in damping down and reconciling the internal disharmony that is endemic within the Navajo psyche.

But as the power of the old medicine men declines, and as they lose their power to reconcile the disparities and schisms of the Navajo character, the Navajo turn increasingly to alcohol to bring about a temporary abatement of the inner struggle. When drunk, the Navajo live out, spasmodically, the oral sadism of their nature, turning it outward in the form of wife abuse or inward in the form of suicidal self-destructiveness. Thus the Navajo live balanced between extremes. They become gluttons toward all forms of experience and use strong drink to heighten their excitement and to release the explosive side of their nature. In their own term, they are *t'sin di'da* (going to extremes), and while they claim that this condition is brought on by the violation of incest taboos, it is in actuality a pervasive trait, one that is barely counterbalanced by their depressive stolidity, another prominent quality. It is as though they had internalized, from the Pueblo Indians that they defeated, qualities of reserve and passive resistance. These inhibitory "Hopi" tendencies are in perpetual war with the wild Apache heritage that shows itself under the sponsorship of alcohol.

Highland Maya Personality: Surgent Features

While the Highland Maya differ from the Navajo in ethnic and cultural terms, they share great similarities with them in the more unconscious, appetitive, and eruptive sectors of personality. Like them, the basic libidinal position is oral-sadistic; they live in a "dog-eat-dog" world, in which human relationships are charged with themes of envy and revenge, as well as fear of powerful, retributive spirits. The unconscious dispositions emerge readily in the projective protocols, in intensive interviews, in the themes of religious life, and in the motifs of drunken behavior. For example, Rorschach tests taken in Chiapas contain many images of animals hunting, being hunted, and being eaten. It is through such imagery that one gets the "dog-eat-dog" picture of people who experience their oral predation as an external, depersonalized threat. That is, these men see in the Rorschach their own oral, "cannibalistic" activities in nonhuman form; this finding suggests that they do not recognize such motivations as part of the furnishings of the self.

The interviews reveal a similar oral underpinning of personality. In answer to my standard question, "What are the things that make you feel good?" men routinely replied that they were content when there was enough to eat for them or their families. By the same token, they were not content when their families lacked food or the money to buy food. For many of them the lost feeling of contentment was restored by drinking *trago* (hard liquor). Many men defined happiness not in terms of human relationships but as sufficiency of food.

Even the more *avanzado* (progressive) men of the village shared these oral concerns. For example, Don Juanito (Juan Perez), perhaps the leading man of the village, an octogenarian of much curiosity and liveliness, questioned me for more than an hour regarding the eating habits of North Americans. However, Don Juanito's "orality" was not inconsistent with a wide-ranging mind and a generous spirit; he craved his dinner and became tense if meals were delayed, but he was also concerned with the oral pleasures of other people. (He once re-

257

marked that progress is a good thing, and he was glad to see progress come to all the world, because progress ensured that all men would have enough to eat.)

Much liquor is drunk in Highland Maya communities, and men who are ordinarily restrained and polite become notably demanding when drunk, and show a more violent face. It can be argued that drunken emotions are not released by booze but are its product; however, since drunken behavior in Chiapas confirms predictions derived from sober behavior, the words of drunken men can be used to estimate "private" aspects of their personality. Thus my first interpreter in Aguacatenango was mild and friendly when sober. Drunk, he alternated between cringing subservience and angry demands, and both themes found expression in a single sentence. Clutching my shirt, he would shout *"Ayúdame mi jefe! Ayúdame papacito!"* ("Help me, my chief! Help me, little father!"). He roughly demands and propitiates in the same breath.

Cultural anthropologists might argue that the independent variable determining Highland Maya character is not some hypothetical oral sadism, but rather their means of livelihood. As subsistence peasants, at the mercy of their arbitrary environment, they must often go hungry. Who then could blame them for having "oral" concerns? But the Lowland Maya are also hard-working, subsistence-level corn farmers who have known famine and are realistically concerned about adequate food supply for themselves and their families; yet they show little evidence of oral fixation at either the conscious or unconscious levels of functioning. The two people, Lowland and Highland, are products not only of their current circumstances but of their vastly different histories. It is the internal record, the charged memory of past environments, that gives special, catastrophic meaning to current stress in the life of the Highland Maya and leads to their special emotional posture. We can view their orality as a persistent feature of character, one that would continue to shape attitude and behavior even under conditions of secure and abundant food supply.

In both the Highland Maya and Navajo we find an uneasy balance or partial truce between inner rages and inner inhibition, registered in an outward show of decorum. The Navajo seem to resolve this tension through extravagant mobility and through the actions of their senior men, who bring to them the presence of the calming, harmony-restoring gods. Finally, particularly in the "detraditionalized" Navajo regions, hard liquor provides an external means for reducing inner tension by turning it into external conflict. Navajo drinking increases in direct proportion to the decline of gerontocracy, as represented by the powerful medicine man of the traditional districts. But the reverse appears to be true among the Highland Maya. In their eyes, strong external control permits and even justifies drunkenness: The likely prospect of punishment constitutes a kind of down payment for the next breach of discipline. The Highland Maya can lose control "safely," in the knowledge that, ultimately, the control function is held in escrow for them by the old men.

Highland Maya Personality: Ego Aspects

The construction of reality—the ways in which reality is experienced, thought about, and understood—is the work of the ego, just as the ego is itself to some degree a product of the reality that it constructs. In Chiapas, the boundary-maintaining, defensive, and cognitive functions of the ego have a decidedly syncretistic quality, a diffuseness of the sort that we have already seen among the Navajo and that is found generally in immature, orally fixated characters. For example, the Highland Maya shows a notable tendency to confuse his own emotions with those external objects that are keyed to strong feelings and associated with their arousal. This is but one aspect of a general ego diffuseness, so that the Highland Maya are not clear where they end, affectively speaking, and the other begins. Thus they often experience their own emotion in the object of that emotion; they externalize responsibility for their desires onto those objects that rouse their desires; they maintain "resonance" theories of causation, believing that strong feeling in one person can cause reciprocal effects in another without intervening action or communication. They cannot easily separate their motives from their acts, and they map their world into categories that are based on what might be called emotional or physiognomic distinctions.

It could be said that this syncretistic mode of thought reflects culture and not personality, that people give the explanations they have been taught to give. However, it appears that this cognitive mode represents more than social convention, for it has been internalized among the Highland Maya and is applied to hitherto unexplained phenomena. For example, an old man in Aguacatenango offered his personal explanation of early mortality:

> I don't know why we are not well formed here. The church is only half completed; it's not whole. We wanted to finish it, but he who finishes it dies. I don't know why the church was never finished. It's not well built.

259

A long time ago there was another church behind where the present one stands, and then there was an earthquake and that temple was ruined. They began to build it again, and they didn't make it the same size. The people were to blame because they didn't have enough money, so they said it wasn't necessary to build it the same length as it was before: "You can leave it just like this." Then the mason who was doing the work said, "All right, I'll do as you say, but then I'll never return to my land; I'll die right here." And that's what really happened; he died right here. That's why I believe that the men don't live very long here.

The syncretistic mode of thought comes through clearly here: A short church *is* a short life; a church finished too soon *is* a man who dies too young. Society may have sponsored this form of thought in the respondent, but this particular explanation represents a private solution, a personal conviction, and a personal relation to experience. It represents ego as much as it represents society.

The externalization of initiative and responsibility is most striking in terms of the use of hard liquor. An Aguacatenango or Zinecanteco man will rarely say that he wants a drink; almost always, as he sees it, drink serves the need of some agent external to him: "My heart wants *trago*," "my tapeworm wants it," "my belly calls for *trago*." Similarly, there is clear confusion of motivation with its object in the statement "The *trago* asks me" ("*Trago me pide*"). The last is reminiscent of the child's favored defense against blame, "He made me do it!"

Liquor does not excuse only the act of drinking; it also excuses the drinker's antisocial behavior when drunk. It is the liquor that bears the responsibility for aggressive and noisy behavior: "Liquor makes their heads go mad; that's why they fight. When they don't drink they are silent and they don't fight. It's the liquor that does it."

The Highland Maya quite concretely feel themselves to be extended into nature and beyond their own skin. For example, they feel that their own existence could be intimately tied up with the fate of their *nagual*, an animal spirit whose life, though passed in the distant mountains, is coterminous with theirs: "If one sees his *nagual*, then he dies, because if the *nagual* gets killed, the man dies as well. The *naguales* get together and fight, then whichever *nagual* dies, his corresponding person dies too. If a man doesn't have a *nagual*, he doesn't die."

The generally concrete and external orientation of the Highland Maya is suggested also by their relative inability to separate action from motivation. Men were not able to conceptually distinguish what they did from what they ought to do or from what they would like to do. For example, men could tell me in great detail how many *cargos* (formal responsibilities for church upkeep and rituals) they had already filled, what they had done in each office, which offices remained to them, and how long it would take to complete their duties to the town. They could not, however, tell me how they felt about the duties that the town imposed on them; they said that they liked to fulfill these duties (even though economic loss is entailed) because they have to fill them: "It is the custom." Similarly, they insist that they like to work, and when I asked them why, they solemnly replied, "Because this is our custom here, to work in the *milpa*." Thus there is little internal division between what might be called an "acting self" and an "intentional self." The Highland Maya are aware of themselves in terms of what they do and in terms of the objects that they deal with, the socially defined roles that they fulfill, and the customs that they obey. They are not aware of a self that chooses among conventional courses of action, that prefers one conventional possibility over another, or that exists distinct from the goals and values of conventional action.

This dispersion of parts of the self into the environment, into action, and into other agents extends to questions of punishment as well as to questions of responsibility. Again, we do not find a split between an acting self and an observing or intentional self; we do not find the split between a self that acts and a self that reviews these actions and criticizes them from some moral perspective. In the experience of these people, punishing agents are located externally.

In this sense, the Highland Maya do not experience an internalized superego and the capacity for independent moral calculation; rather they experience the precursors of the superego, for example, a fear of external punishment linked to specific situations and specific misbehaviors. As the following excerpt suggests, responsibility for both sin and punishment is in the hands of bad people "out there":

> When we drink liquor, we aren't all good people; some are bad men and others good. The good ones just pass the day happily and the bad ones come around and they see two or three of us are happy. He comes around and we begin to fight. But then again, when we are sober we begin to feel sad. We think about what we said and why we began to fight. We strike one another; that's why we grow sad.

The Highland Maya experience not internal guilt but punitive outer presences; they are by the same token very concerned with the destructive power of agents external to themselves. For example, when dealing with the TAT, the Highland Maya constantly introduce categories and distinctions based on size and power, while the Lowland Maya, looking at the same cards, see wholes and complete gestalts. Thus, responding to the Bats and Man scene, Lowland Maya see a swarm of bats flying over a recumbent man, while the Highland Maya are not satisfied with the encompassing generalization "swarm," but instead concern themselves with the pecking order among the individual bats: "This is the biggest one, and he is taking food away from the smaller one."

This concern with gradients of power and strength, and of the envy that travels between these levels, directly affects the relationship between men and women. Though men marry women, cohabit with them, and have children by them, the sexes do not share an intimate life. In fact, Highland Maya men seem happiest when they are away from women, working in the *finca*, in the *milpa*, or drinking with the boys. Women are smaller and weaker than men; as a result, men seem to discover in women the oral and predatory yearnings that they themselves covertly bear toward men of superior status and power. In effect, men see in women the castrating wishes that they bear toward men superior to themselves in possessions and strength. Thus, in local myths and folk tales, women are seen as dangerous beings who would drain men, rob them of their strength and flesh; hence they must be controlled, "kept in their place." It is as if the men say about women, "If I were castrated, I would hate the men because they have what I lack. Thus women are a threat to me."

Certainly, the women of these villages exhibit a great deal of shame, especially as compared to the self-confident Mayan matriarchs of the Yucatan, and they are quick to cover their faces with their ever-present shawls, as if to hide dirty or evil things. It is as if the women have accepted the men's judgment about them, as if the women (perhaps shamed by their own "castration") confirm the male projection and admit that men are correct in thinking of them as bad or untrustworthy. In this regard, the sexual parts of a woman are considered to be so horrifying that even the devil would be frightened of them. Men of Tenejapa sometimes insist that their wives accompany them into the mountains, the haunts of *pukuhes* (demons). The wife has but to raise her skirts to show her genitalia and the demon will flee.

Druze Personality: Surgent Features

Though many Druze are illiterate, as a people they are not in the technical sense preliterate; though peasants, they are in touch with Islamic civilization, they value education, and many of my informants had received early instruction in written Arabic from the *khatib*, the traditional itinerant scholar of the Levant. In addition to this proto-literacy, individual Druze are determinedly rational, even intellec-tualizing, and capable of a highly sophisticated moral calculus. Thus an elderly Druze gives these reasons for refusing the social security payments offered by the government of Israel: "Social Security payments come from taxes. Taxes represent money that is taken against the will of the taxpayer and ultimately by the force of the government. Therefore, for me to accept social security would be like accepting stolen money, money unwillingly given, and I cannot do this." In this comment we hear an essential note of Druze character. If the Navajo and the Highland Maya are characterized by oral dependency, and by the frustrations and rages contingent on that stance, the Druze are characterized by a rigid counterdependency; unable to take the goodwill of the majority for granted, the Druze have learned as a group to mainly trust themselves—the courage, strength, and wisdom of the Druze people—and the individual Druze tends to reproduce in his character the counterdependent, ferociously self-reliant posture of the culture as a whole. Thus the Druze would allow themselves to depend only on those resources that are unequivocally under their personal control: their own capacities or the resources that they themselves have cultivated, such as the produce of their fields and the strength of their sons. In effect, they trust very little that comes from outside themselves as gift or gratuity, and this mistrust for what is not under self-control extends even to their emotional life. Thus, as befits a surviving minority, individual Druze are extremely stub-born and refractory, not only against coercion from another's will, but also against coercion from their own willfulness, their own spon-taneous emotions. They do not trust those parts of the self that are not clearly marked with their own brand, which are not called up by or controlled by the personal will.

Druze Personality: Ego Aspects

Accordingly, not allowing themselves to give way to excitement, even illiterate Druze peasants give priority to rationality over emotionality. In effect, they value the mental resources that they have created for themselves over the emotions that have been "foisted" on them. As part of their self-reliance, they must deny any needfulness (except in regard to Allah) and must hold themselves ever in the position of the giver rather than the receiver. They are unfailingly and even aggressively hospitable; the entire Druze household is mobilized toward the reception and nurture of the guest. If the male head of the house is not at home, the guest will be met by a son, who will attempt to hold him with food, with coffee, with conversation, and with promises of his father's imminent return. And if the household males are not present, the guest will be met by the women, usually kept out of sight in the inner recesses of the home, who offer the same comforts and make the same promises. The worst insult that one Druze can give another is to call him a "hotel keeper," meaning that he accepts money from his guests in exchange for hospitality.

But while Druze men stifle their own dependencies, they are fiercely loyal and unflaggingly nurturant toward their dependents—their children and their wives. The Druze man can cherish and tend in his wife and young children the dependencies that are anathema within himself. This projective pattern might bespeak a collective neurosis, but in the Druze case it guarantees reliable and devoted parenting. The fathers are protective and warm, and the women, feeling secure in their husbands' affections, tend to be loving mothers, who pass on to their children their own sense of assurance and protection. As a result, the Druze rear very attractive and sturdy children, responsible and obedient, though not at all lacking in spirit and zest. Again, as among the Lowland Maya, it is rare to find Druze children, particularly those from the same *hamula* (clan), in contention with each other.

This equanimity is preserved within and among all social levels: among children, and among adults within the same family and the same clan. Given the sense of reliable inner controls, as well as the norm of hospitality, the Westerner who penetrates the Druze community can feel far more secure in their remote villages than in his native city.

Despite their reserve, their resistance to coercion, and their mistrust of strong feelings, the Druze make excellent subjects for my kind of research. To some degree, the norm of hospitality neutralizes the factors that work against effective interviewing. Having accepted you into his home, the Druze subject is honor-bound to comply with your request for a full interview, and once engaged, his capacities for introspection and self-observation come into play. The resulting exchange is constructive for the researcher and even therapeutic for the subject.

Lowland Maya Personality: Surgent Aspects

While the Lowland Maya bear a resemblance, in terms of basic personality, to the Druze highlanders, and even to the burghers of Kansas City, they are notably different from the Navajo and Highland Maya Indian groups. Whereas the latter are given to excess, heavy drinking, and an unstable balance between control and impulsivity, the Lowland Maya are notable for their high degree of inner balance, with moderation, in appetite as well as in controls on appetite, as a central theme of Yucatecan character.

Thus the Maya of the Yucatan are distinguished from the other Indian groups of our sample not only by their special terrain, language, and practices but also by a special strand in their psychosexual makeup. For want of a better term, we will call this special character motif "anal retentive." This usage is not a piece of Freudian mythologizing; the Mayan people made this term more real for me than did members of my own society, where the term was first coined. The Maya show all the classic stigmata of this character type: They are tremendously concerned with thrift (and there are abounding rumors of buried treasure); despite their relative poverty, they are compulsively clean, washing themselves as soon as they return from the fields; they are orderly with possessions and with time, always managing to be engaged in some productive activity; and their disciplined industry is internally rather than externally driven, requiring no heavy superstructure of authority or outward coercion. But the practice that is most emblematic of the anal character is found, naturally enough, in the area of toilet training: Mayan parents report that their children are continent for both feces and urine by the age of six months, an impossible achievement by the norms of American developmental psychology. Yet we never saw a soiled child in the Yucatan; and though six-year-old girls carry their youngest siblings undiapered, on the hip, in the traditional *hetzmek* position, yet we never saw a soiled dress.

Young parents were ashamed for a six-month-old hammock wetter and would plead that the child was properly continent by day, with only occasional lapses by night.

We did not learn how this extremely early training was accomplished, but it clearly involves neither harsh punishment nor intimidation. One senses that rigorous training begins as soon as the child becomes responsively "human" and is accomplished by means of quiet, stubborn persistence on the part of the parents: The child is put on the pot again and again until he or she gets the essential idea. This training method is reminiscent of the general tenor of Mayan socializing efforts, and it also reflects the quiet, persistent way in which the Mayans conveyed their expectations of proper behavior to me and my family during our stay in their village. Judging from our "childhood" in their community, Mayan training, whether for the toilet or for more public behavior, is accomplished through pressure applied subtly and allusively, so that the child can neither identify the parental coercion as such nor muster himself to rebel against it. The Mayan child may find it hard, from the first, to sort out parental pressures from emerging self-requirements. When demands are calmly but persistently laid down long before trustworthy self-other boundaries have been established, they may be experienced as part of the very texture of self, and not as distinct, externally located intrusions at variance with the self. The Mayan parents' nurturance, quiet warmth, and evident readiness to sacrifice for their children also make it harder for their offspring to oppose themselves to parental requirements, and may foster their identification with controls that originate in parents and in culture.

Maya Lowland Personality: Ego Aspects

Despite their rejection of sensuality and of extravagant action, the Lowland Maya are neither determinedly puritanical nor joyless; despite their stringent emphasis on work and thrift, they are neither dour nor sour. They joke, gossip, and laugh; their facial expressions are warm and sometimes even merry. They like sports, and many men have a delightful though understated sense of humor. As regards life's pleasures, the Mayan goal appears to be moderation rather than abstinence. All the pleasures—of carnival, of the dance, of cuisine, of drinking, of sex, of sport, of the hunt—are represented in the village, but always in moderate degree.

In short, the orderly conduct of affairs in the Yucatecan village seems to be based on an internal and inflexible readiness to accept self-imposed restraint and discipline. The readiness to accept common standards, as well as the standards themselves, are shared with remarkable unanimity by most adult members of the society. For example, Mayan TAT results are striking for their degree of what might be called "implicit conformity"; individual respondents did not know how other villagers had interpreted the TAT cards, but most respondents would, in their private and spontaneous constructions, closely approximate the consensus of their peers concerning the card "story." There were, of course, variations in interpretation, but these fall within a narrow thematic and structural band. Even the so-called misinterpretations of unstructured cards showed a remarkable uniformity; for example, most Yucatecans saw in the Desert Scene TAT card not a desert or a field but an ocean. In terms of what the artist and researcher intended, this is a clear misinterpretation. What is remarkable is the repeated occurrence of the same notable misinterpretation from villager to villager.

A transcultural regularity begins to emerge. When the responsibility for instigating behavior and controlling it is not internalized (among the Navajo, and the Highland Maya), dominant men are charged with the power and the responsibility for disciplining the wayward young. But when controls are internal, trustworthy even in the absence of external authority, then we find, as among the Yucatan Maya, that the aged may have no social power. They may be accorded respect and are exempted from the usual demands for productivity, but they are not granted a heavy hand to punish the transgressions of the young. It may be that the welfare of the aged rests on two independent bases: the fear and awe that they can inspire, and the gratitude that, as parents, they have engendered in their offspring.

The Horse and Men Card

The Horse and Men card allows us to pinpoint more exactly the social shaping of the aggressive motives in middle and later life. The depicted scene (fig. A–1) shows two men, one indigenous and the other from the dominant, majority culture—a white man in the American Indian

FIGURE A–1
The Horse and Men Card (Druze Version)

case, a Ladino in the Mayan case, and a European (possibly an Israeli Jew) in the Druze case—in an intense debate, possibly over a horse, whose bridle is held by the indigene. An important card issue has to do with the stranger, particularly the powerful stranger: how to approach him, how to cope with his power, and how to keep one's possessions (in this case, the horse) from his grasp. But the more significant, central card pull has to do with aggressive tension between males and with masculine competition over some desired object, some token perhaps, of virile manhood. Again, we can regard these linked issues—the male struggle, the dangerous contact with the powerful stranger—as the dynamic background of most responses to this card, including even those responses that do not manifestly refer to such card issues. Thus we assume that the respondent who does not openly mention these matters is unconsciously aware of them but unconsciously avoiding them. Our right to apply this logic comes from the fact that 62 percent of all subjects who responded to this card addressed themselves explicitly to its aggressive possibilities and thereby defined its latent stimulus demand.

The Horse and Men Card

The criteria for sorting Horse and Men stories according to mastery types are as follows:

Active Mastery

Grouped here are all those stories (AM_1) in which the combative possibilities of the card are faced head-on and are even amplified, though without violation of the objective "realities" of the card. In some cases, the two men are seen as struggling directly and even physically for possession of the horse. Their struggle is not tempered and contained within formal contests, such as bargaining. However, the battle does not lead to frightening or permanently destructive outcomes for either participant, and the respondent tells his story with some zest or appreciation of the mayhem that he discovers in the card.

Other stories under this general heading (AM_2) reflect some social buffering of the intermale contest. The protagonists are hotly engaged in a commercial struggle, in bargaining rather than physical assault. Here the question is not which man will possess the horse but the amount of compensation that the owner will get in exchange for it. The struggle is deflected from the tangible goal, the phallic horse, to the more abstract economic value that the horse represents.

Qualified Active Mastery

In these stories, though the perceived action fits the stimulus realities, the aggressive component of the story has escaped from its usual moral and social inhibitions, and leads to bad consequences. In these stories, a thief, who may or may not be successful, attempts to take the horse by force from its rightful owner. In either outcome, aggression is seen as suspect, dangerous. If the thief is successful, moral norms, the bases of the social contract, have been violated; if the theft is prevented, the thwarted thief is usually punished for his violation. In either case, unchecked aggression carries an inordinately high price, one that neither society nor the criminal can afford to pay.

Bimodal Mastery

In these stories, while conflict possibilities may be acknowledged, they are also distanced and contained. For some (QAM_1), while there has been a commercial contest, a bargaining session, the story centers

269

on the end point of the exchanges, when an agreement is reached. The contest is over and is being replaced by amicable agreement. In his perception, the respondent substitutes the possible (but not depicted) pacific resolution in place of the immediate, ongoing contest commonly suggested by the card.

In other stories of this mingled active-passive variety, the tactic of distancing and isolating aggression is more clearly elaborated. Thus the conflict is not waged full-blown, between two participants: Aggression is assigned to only one figure, while the other man is defined as more passive, subordinate, or amiable (QAM_2). In effect, the subject distributes his wish to attack into one figure and his wish to retreat, to passively avoid trouble, into the other protagonist. This subtype also includes stories in which one man attacks while the other placates him; those in which some violent movement of the horse has touched off a quarrel between the two men; those in which the issue of dominance has already been settled (so that one protagonist arrogantly commands the other); and stories that are undecided between conflict and collaboration, in that they start with one such motif and end with its opposite (QAM_3).

Passive Mastery (Sensual)

In stories that qualify for this mastery position, themes of easy affiliation have replaced the themes of struggle or potential struggle between the two men. Instead of fighting or trying to advance their own interests, the men are positively helpful to each other: One man gives travel directions to the other; the owner of the horse instructs the foreigner in horse care or loans him his steed. In these detoxified versions, the upraised arms of the two men connote warmth, a desire to embrace, rather than threat or combat.

Passive Mastery (Constriction)

Here, criteria for inclusion call for stories in which there is no mention of aggression between the two men. Aggression may be outside of the human dyad and is usually lodged in the horse. The horse is wild, and the men collaborate in taming it, or the horse is startled by some strange sight or sound (PM_1).

Some stories stress formal or neutral rather than intense and fierce contact between the two men. For example, the subject interprets the upraised arms of the two men to mean that they are saluting rather than hitting or threatening each other. Along similar lines, though potentially divisive ethnic or social differences between the two figures are noted, these do not lead to a conflict: Two acknowledged strangers talk, greet each other on the road, etc. (PM_2). In other entries the stimulus is described but not interpreted: The subject notes only that there are two men and a horse. There is minimal if any introduction of plot, affect, or action (PM_3).

Magical Mastery

In these stories, the aggressive or benign possibilities suggested by the card have been elaborated and escalated beyond the reasonable limits allowed by the stimulus. One man menaces the other with a gun or one combatant has already murdered the other. Conversely the figures are presented as impotent rather than armed: "This Druze man has a sick arm; he is raising it, showing it to the other" (MM_1). On the benign side (MM_2), the two men are not only friendly, but positively loving: "They are very happy together" (Lowland Maya). Or one figure is seen as an angel that blesses the other. But in all these cases, whether the protagonists are loving, armed, or crippled, the basic features of the card have been misperceived, to emphasize, on the one hand, mutilation or murder, or, on the other, the total denial of such fearsome possibilities.

The distribution of reponses to the Horse and Men card, by age, mastery style, and culture, are given in table A–1. Figure A–2 presents the mastery profiles generated from the stories given by younger respondents (aged 35 to 49) and by older respondents (60+) from four societies: the Navajo, the Druze, and the two Mayan groups.

As table A–1 indicates, the mastery themes do not strongly discriminate the age groups: The statistical test of the mastery stage by age shows a real but not pronounced trend in support of the research hypotheses. However, the age trends are in line with my predictions and with the more dramatic trends in the data from the Rope Climber and other cards. As figure A–2 shows, the younger men achieve their highest scores on Active Mastery and on Qualified Active Mastery. Their unexpectedly strong showing on Passive Mastery (Constricted) suggests that their fund of available aggression may create tensions that they try to manage through some inhibition of fantasy and action. The conclusions to be drawn from the mastery profile generated from the older men's data are that the emphasis has shifted more strongly

271

FIGURE A-2

The Horse and Men Card: Comparison of Mastery Distributions:
All Younger (35+) and Older (60+) Male Subjects

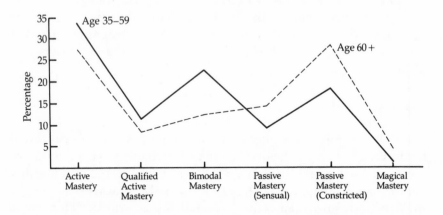

in the direction of inhibition, passivity, and avoidance of conflict. Thus the older men always achieve rankings on Active and Qualified Active Mastery that are lower than or at best equal to those achieved by the younger men, and the older men rank higher than the younger men on all the versions of Passive Mastery. Their most notable gain is on Passive Mastery (Sensual), suggesting their need to avoid occasions of anger and to adopt instead the role of peacemaker. Their fear of conflict is suggested by the fact that Passive Mastery (Constricted) is their favored posture, and they may evade the fear refracted in this style through adopting the conciliatory, helpful approach to potential enemies. It is as if they tell possible rivals, "I do not want anything from you; rather, I want to help you. Clearly then, you have nothing to fear from me (nor I from you)." The Rope Climber card showed us that older men tend to dissemble their aggression through projection, through discovering it as an external threat in others; the analysis of the Horse and Men card suggests another defensive strategy: denial. Besides projecting their aggression, older men hide from it and conceal it from others through asserting its opposite: "I wish to help, not to hurt."

The Horse and Men card was readministered to the original panel of Galilean Druze subjects at $Time_2$. The results of the $Time_1$ versus $Time_2$ comparison are shown in figure A-3. In this case, the hypothesis that age changes in card perception within individuals would replicate age differences between cohorts, as found in the original data, was

TABLE A–1

The Horse and Men Card: Distribution of Responses by Age, Mastery Style, and Culture

Mastery Style	Subjects	Age				Total
		35–49	50–59	60–69	70+	
Active Mastery (AM₁, AM₂)	Navajo	19	6	13	10	
	Druze	16 37 (.34)ª	16 25 (.31)	13 29 (.33)	5 15 (.22)	106
	Lowland Maya	1	1	3	0	
	Highland Maya	1	2	0	0	
Qualified Active Mastery (QAM₁, QAM₂)	Navajo	2	6	4	0	
	Druze	8 13 (.12)	3 13 (.16)	10 14 (.16)	1 1 (.01)	41
	Lowland Maya	1	2	0	0	
	Highland Maya	2	2	0	0	
Bimodal Mastery (QAM₃)	Navajo	6	5	5	2	
	Druze	11 25 (.23)	4 15 (.18)	3 12 (.14)	4 9 (.13)	61
	Lowland Maya	6	5	4	3	
	Highland Maya	2	1	0	0	
Passive Mastery (Sensual) (PM₄)	Navajo	2	3	3	5	
	Druze	7 11 (.10)	4 10 (.12)	5 12 (.14)	2 12 (.18)	45
	Lowland Maya	1	3	2	3	
	Highland Maya	1	0	2	2	
Passive Mastery (Constricted) (PM₁, PM₂, PM₃)	Navajo	1	2	2	2	
	Druze	6 20 (.19)	6 17 (.21)	7 18 (.20)	12 27 (.40)	82
	Lowland Maya	5	5	4	6	
	Highland Maya	8	4	5	6	
Magical Mastery (MM₁, MM₂)	Navajo	0	1	0	1	
	Druze	1 2 (.02)	0 2 (.02)	1 3 (.03)	3 4 (.06)	11
	Lowland Maya	0	1	2	0	
	Highland Maya	1	0	0	0	
TOTAL		108	82	88	68	346

ª The chi-square test was applied to the cell totals of the combined mastery types: Active, Qualified, Bimodal, Passive, and Magical Mastery. The probability that this condensed distribution occurred by chance is less than .025.

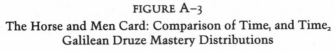

FIGURE A-3

The Horse and Men Card: Comparison of Time₁ and Time₂
Galilean Druze Mastery Distributions

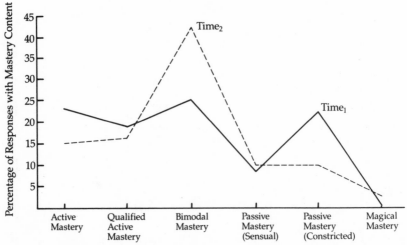

not borne out. As figure A-3 indicates, there is, as predicted, a clear decline in Active Mastery responses by Time₂, but there is also a strong decline, against my prediction, in Passive Mastery responses. The longitudinal shift is not from Active to Passive Mastery, but away from both Active and Passive Mastery toward Bimodal Mastery. The statistical difference between the Time₁ and Time₂ distributions is significant, although the variance is mainly accounted for by the notable Time₂ peak on Bimodal Mastery. Thus, while the developmental hypothesis has not been greatly supported by the longitudinal data, these data are marginally in line with the predictions from that hypothesis, in that the central tendency, over time, is to move away from an unambiguous toward a more equivocal stance in terms of intermale combativeness.

Aggression-eliciting TAT Cards:
Intercultural Analyses

The aggregate results of TAT administration, from all respondents across all cultures, as presented and discussed in the text. In this section, I will discuss two cards of aggressive implication on a more finely tuned, culture-by-culture basis. The mastery codes and basic, aggregate distributions for the Rope Climber card have already been reviewed in the main text; here I will review the cultural variations on the main card themes. The Male Authority card has not been discussed in the text; in this section, I will present the card as a stimulus, and discuss the mastery codes, the aggregate results, and the intercultural variations.

The Rope Climber Card

Generally speaking, when we compare the mastery profiles of younger and older men in all the cultures for which we have Rope Climber data, we find a pattern of age differences that is in each case similar to the one produced by the aggregate data for that card (see figure 3.1).

Kansas City. In these data younger respondents have a great lead over older respondents in the production of stories classified under Active Mastery$_1$ and Active Mastery$_3$, those having to do, respectively, with the pleasures and perils of competitive achievement. However, older respondents hold their own very well and even edge out the younger respondents on those mastery positions involving modulated prosocial and domesticated versions of Active Mastery and Qualified Active Mastery. Older American men are much more apt than younger American men to experience their own aggression as an external menace rather than as an internal, though sometimes problematic, resource. The older men achieve high rank on Passive Mastery$_1$ (ten stories of this sort as against none for the young men) and on Magical Mastery$_1$. Again, for the older men, aggression loses its familiar phrasing, as a reasonable though sometimes troublesome component of the self, to become an exterior threat or pressure. However, when the older Kansas City men project their aggression, it is not via the rather primitive mode registered in MM$_1$ (the same mode favored by older men from our sample of preliterate societies). For elderly Kansas City men, aggression does not regularly become an aspect of the alien, supernatural "stranger"; rather, it remains within the human realm, even though it is not a welcome part of the social landscape.

Navajo. The Navajo age/mastery profiles also map the cohort differences that are regularly found in the larger cross-cultural sample. The younger men have the usual clear advantage on the indicators of open aggression—AM$_1$, AM$_2$, AM$_3$, and PM$_1$. Younger Navajo men score highest on AM$_2$—not the best index of free, competitive motives;

however, in the Navajo case, AM_2 stories mainly have to do with the theme of physical exercise, in which the hero prepares for future contests. Thus, for the Navajo, AM_2 is actually associated with competition and the exercise of power against others.

Druze. The "classic" age pattern for this card is even more clearly graphed by the Druze data, although here the younger men score higher than their seniors on all but one measure of Active and Bimodal Mastery. Again, the younger men first give precedence to the older men only at that point in the mastery spectrum, PM_2, at which aggression—at least in its pragmatic and "reasonable" forms—ceases to be an issue. Older men produce a characteristic surge on MM_1, signaling the prevalence of dissociated, "eerie" aggression in their fantasies. Again, as with the Navajo, the aggression that was available to, or at least comprehensible to, the younger men is disowned by the older men but then returns to become the stuff of their nightmares or of their religious obsessions. As contrasted with the Kansas City men and the Navajo, there is a general rightward shift in the profiles of both Druze age cohorts: The younger men score lower on controlled aggression, and the older men score higher on Magical Mastery than do those of the other groups.

Maya. Somewhat different patterns are graphed by the data from the two Mayan groups. Though in both the Lowland and the Highland Maya cases there is a statistically significant age-related drift toward a more quiescent orientation, both groups seem less expressive of or less sensitive to the aggressive and power implications of the Rope Climber card. Both Mayan groups show no entries for AM_1, and, while a small percentage of the Highland Maya responses are grouped under AM_2, they give priority, by six percentage points, to AM_3. They are more cognizant of the destructive rather than the constructive and pleasurable aspects of their own aggression. (For the non-Mayan societies, AM_3 ranks either below AM_2 or is at best equal to it.) Thus, for the Maya, whether Lowland or Highland, personal assertion is suspect, haloed with danger and therefore subject to much inhibition. This caution is particularly manifest in the stories of the Lowland Maya: In almost every case, regardless of the climber's activity or his goals, the story ends with the rope breaking and the climber falling to his death.

In sum, the younger Maya project vigorous activity only in connection with limited or consummatory rather than large-scale and competitive goals: Within the Active Mastery spectrum, they find their voice in QAM_2 and QAM_3, the registers of muted, aim-inhibited assertiveness. Indeed, the mastery profiles traced by the data from the younger Mayan men match closely those produced by the older White North Americans, Navajo, and Druze: low on AM, high on QAM, and high on MM_1. However, despite the relatively senescent profile of the younger Maya, there is an independent age effect: The mastery profile of the Highland and Lowland men falls even further to the right on the mastery spectrum. Instead of peaking on MM_1—as is normal for the older men of the non-Mayan groups—the older Mexican Indians are best typified by PM_4, a dependent, self-indulgent version of the Rope Climber. For them, he is a man who rather irresponsibly plays and diverts himself with the rope. This finding in regard to the older Mayan men supports our interpretation of the meaning of the age shift to MM_1 as found in the other cultures: Because the younger Maya do not flaunt their aggression, the older men are not haunted by it. Perhaps in societies, such as the Maya, in which there is little sponsorship for self-assertion, the younger men do not have to manage their aggression through bold activity, and the older men do not have to manage it via the paranoid projection of MM_1.

A hypothesis begins to take shape: Societies that promote the aggression of younger men may have to—paradoxically—support the religiosity of older men. Those men who have been trained early to seek, develop, and relish their own power will continue to seek it in later life as well, through alliances with the supernatural. The routes to such sacred power would be made available to them via the religious life and its rituals.

Thus the men of "aggressive" cultures may move toward a paranoid religiosity in later life, in contrast to the men of "counteraggressive" cultures, who move to a more receptive, worshipful stance. But the major and general trend, across all cultures, seems to be from Active Mastery to passivity. The cross-cultural data underlying these conclusions may be found in tables A–2 and A–3, and in figure A–4.

The Male Authority Card

Controlled aggression is the basic fuel of all hierarchical social systems. Accordingly, if the male stance toward energy and aggression typically changes over the life span, we would expect corresponding age changes in male attitudes toward masculine authority hierarchies as well. This assumption, that men, as they age, would revise their conceptions of dominance and hierarchic power, was tested with the Male Authority card, a stimulus used at all sites except Kansas City.

TABLE A-2

The Rope Climber Card: Distribution of Responses by Age, Mastery Style, and Culture

Mastery Style	Subjects		Age			
		35–49	50–59	60–69	70+	Total
Active Mastery (AM_1, AM_2)	Kansas City	9	16	7	1	
	Navajo	13	6	8	4	
	Druze	11 34 (.25)[a]	5 30 (.19)	3 18 (.14)	3 8 (.11)	90
	Lowland Maya	0	2	0	0	
	Highland Maya	1	1	0	0	
Qualified Active Mastery (QAM_1, QAM_2)	Kansas City	6	17	3	1	
	Navajo	2	6	5	3	
	Druze	11 29 (.21)	12 44 (.28)	8 20 (.15)	5 10 (.14)	103
	Lowland Maya	5	6	3	1	
	Highland Maya	5	3	1	0	
Bimodal Mastery (QAM_3)	Kansas City	8	20	13	0	
	Navajo	3	2	6	3	
	Druze	10 27 (.19)	7 37 (.24)	6 30 (.22)	4 11 (.15)	105
	Lowland Maya	3	5	5	2	
	Highland Maya	3	3	0	2	

Mastery Type	Ethnic Group					
Passive Mastery (Sensual) (PM$_4$)	Kansas City	0	3	1	0	
	Navajo	3	2	1	3	
	Druze	1 6 (.04)	1 11 (.07)	0 10 (.07)	2 9 (.13)	36
	Lowland Maya	1	3	4	4	
	Highland Maya	1	2	4	0	
Passive Mastery (Constricted) (PM$_1$, PM$_2$, PM$_3$)	Kansas City	4	10	12	0	
	Navajo	8	7	10	5	
	Druze	7 27 (.20)	3 23 (.15)	5 30 (.23)	5 18 (.25)	98
	Lowland Maya	5	1	2	3	
	Highland Maya	3	2	1	5	
Magical Mastery (MM$_1$, MM$_2$)	Kansas City	2	6	5	0	
	Navajo	3	1	7	3	
	Druze	6 15 (.11)	1 11 (.07)	12 25 (.19)	10 16 (.22)	67
	Lowland Maya	2	2	0	2	
	Highland Maya	2	1	1	1	
TOTAL		138	156	133	72	499

[a] The chi-square test was applied to the cell totals. The probability that this distribution occurred by chance is less than .005.

TABLE A–3
The Rope Climber Card: Comparison of $Time_1$ and $Time_2$ Responses Given by the Same Druze and Navajo Subjects

Subjects	Active Mastery	Qualified Active Mastery	Bimodal Mastery	Passive Mastery (Sensual)	Passive Mastery (Constricted)	Magical Mastery	Total
Navajo: $Time_1$	20	19	6	4	19	4	72
Navajo: $Time_2$	13	11	6	6	25	11	72
Druze (Galilee and Golan): $Time_1$	18	20	24	1	14	16	93
Druze (Galilee and Golan): $Time_2$	14	12	30	4	13	20	93
Combined Navajo and Druze							
All Navajo and Druze: $Time_1$	38 (.23)[a]	39 (.24)	30 (.18)	5 (.03)	33 (.20)	20 (.12)	165
All Navajo and Druze: $Time_2$	27 (.16)	23 (.18)	36 (.22)	10 (.06)	38 (.23)	31 (.19)	165

[a] The chi-square test was applied to the category totals. The probability that the bimodal (all $Time_1$ vs. all $Time_2$) distribution occurred by chance is less than .10. However, when mastery categories are collapsed from 6 to 4 (Active Mastery, Bimodal Mastery, Passive Mastery, and Magical Mastery) the probability that the resulting distribution occurred by chance is less than .025.

FIGURE A-4

The Rope Climber Card: Comparison of Time₁ and Time₂
Combined Navajo and Druze Mastery Distributions

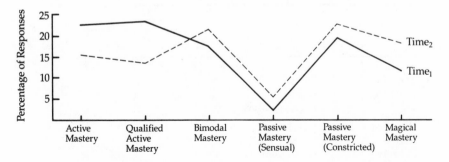

Because it was meant to portray conventional age-graded relationships, it was redrawn for each site, in conformity with local dress styles. The Druze version of this card is shown in the text as figure 6.4.

This card speaks of hierarchies of physical as well as social power. The man is physically larger than the two younger individuals, he is older, and he approaches them in an active and possibly commanding manner. Approximately 55 percent of all subjects agree that the card has to do with authority in that they see the older man in some superior position vis-à-vis the two younger individuals. Some see him as giving orders, others as a resource person who advises without commanding; but in most cases, the older man has power, knowledge, or both, that the younger individuals lack and that he conveys to them. Accordingly, individual responses—even those that on the manifest level ignore the authority issue—can be seen as answers to the implicit question posed by the card: "Where do you stand in terms of authoritative strivings in yourself and others?" Thus, in their formal structure and content, individual responses can be viewed as metaphors of the respondent's subjective relationship to his own achieved prestige or his own strivings toward prestige and authority. The subject's response gives us clues as to whether he is comfortable with authority or status strivings, is afraid of such strivings, or denies the status and social power that he has already achieved. Consistent with the practice followed with the other TAT cards, individual responses were coded, blind for age, as to whether they represented active, passive, or magical management of the authority-power issues dramatized by the card.

The various mastery positions and the criteria that relate particular card themes to the various mastery types are described below.

281

TABLE A–4

The Male Authority Card: Distribution of Responses by Age,
Mastery Type, and Culture

Mastery Type	Culture	Age 35–59		60+		Total
Active	Navajo	25	70 (.36)ª	18	42 (.26)	
Mastery	Druze	35		16		
(AM₁, AM₂)	Lowland Maya	3		6		
	Highland Maya	7		2		
Qualified	Navajo	15	37 (.19)	15	29 (.17)	
Active	Druze	16		11		
Mastery	Lowland Maya	1		1		
(QAM₁, QAM₂)	Highland Maya	5		2		
Bimodal	Navajo	2	8 (.04)	0	8 (.05)	
Mastery	Druze	4		6		
(QAM₃)	Lowland Maya	2		2		
	Highland Maya	0		0		
Passive	Navajo	0	17 (.09)	0	15 (.10)	
Mastery:	Druze	5		10		
Sensual	Lowland Maya	11		5		
(PM₄)	Highland Maya	1		0		
Passive	Navajo	11	51 (.26)	7	52 (.34)	
Mastery:	Druze	12		20		
Constricted	Lowland Maya	18		18		
(PM₁, PM₂, PM₃)	Highland Maya	10		7		
Magical	Navajo	4	12 (.06)	5	13 (.08)	
Mastery	Druze	3		4		
(MM₁, MM₂)	Lowland Maya	5		3		
	Highland Maya	0		1		
TOTALS			195		159	354

ª The chi-square test results indicate that this distribution is not statistically significant and could have occurred by chance.

Before considering these, I should first point out that the overall, transcultural distributions (see table A–4) do not show a statistically significant age shift toward the less-aggressive presentations of male authority. Nevertheless, the Druze, the group most invested in unquestioned patriarchy, do demonstrate, at a significant level, the predicted age differences in cross-sectional data, as well as the predicted shifts, toward less aggressive versions of male authority and interaction, in longitudinal data. Accordingly, I will discuss the transcultural data for this card, while paying special attention to the Druze (Galilee and Golan) results.

The mastery criteria for the Male Authority card are as follows:

Active Mastery. These stories concern the exercise of male authority, sometimes in the face of opposition from the young, in the service of practical economic and educational goals.

In the first cluster (AM_1), the older man orders the younger individuals to perform some task or to go to school, and one or both of the young men oppose him, either with arguments or by deliberate inattention. However, despite this resistance, the older man is clearly dominant. Accordingly, in the stories grouped here, patriarchal authority, bluntly exercised, prevails even when it is opposed. Energy and assertion are found in all male agents, but productive ends are served despite conflict among them, and the mature authority figure remains in charge.

In the second cluster (AM_2), the older man is in control, and his authority is unchallenged, as he directs an obedient younger group, ordering them to work or to school. Again, both generations are energetic, but the energy of the young men is funneled toward production or education, while the older man's energy takes a more personally assertive form and is directed toward controlling the behavior of the younger men.

Qualified Active Mastery. In these stories, the exercise of paternal authority leads to a destructive outcome, or, conversely, it has very little effect, for bad or for good. In the first cluster (QAM_1), the older man is in charge and the younger men recognize this, but the aggression that is part of his authority has gotten out of control, with frightening and potentially destructive consequences for the younger men. The older man is enraged; he shouts reprimands at the younger individuals, and at least one of them cowers away from him in fear. In these stories, aggressive authority does not lead to unqualified good outcomes.

Under the QAM_2 category, the older man is presented as a resource person rather than an authority; whatever legitimacy he has rests on his role as a moral mentor or practical guide. He is not powerful, merely "good" or useful. The older man instructs the younger group on the "good life," or he offers unspecified advice. In a few cases, he gives travel directions. There is no assurance that his advice or direction will lead to positive future outcomes.

Bimodal Mastery. In these stories the older man's authority is purely formal, or it is called into question by the respondent himself:

"He thinks that they respect him, and they pretend to because of his age" (Druze). In this case, the older man receives a purely cosmetic respect because of the ascribed status that comes with age, not because of his intrinsic strengths of wisdom or command.

Also clustered here are stories in which the older man's powers or authority are mentioned by the subject only after much prodding, or in which these qualities are at first introduced into the story but are subsequently withdrawn by the subject. In these cases, the older man is first described as a powerful figure and then as a beggar.

Passive Mastery (Sensual). Here, the older man is neither a provider nor an authority. He does not satisfy the younger group's needs for guidance and control, and the power-authority dimension has been replaced in these stories by the older man's need for intimacy and succor. Thus the older man asks directions from the younger group or begs them for food and money. (In these "beggar" stories, the taller figure of the younger group is usually seen as a woman who hears the older man's plea; she decides whether to help him or deny him. In effect, the power position has been reversed; the younger, now feminized, group is in charge and the older man is supplicant.)

Passive Mastery (Constricted). Conflict and aggression have been deleted from all these stories; instead, the focus is on external forces that oppress all the men equally, or on formal, egalitarian relations within the group. There are no authorities, young or old.

The PM_1 cluster accommodates all stories in which tension or conflict do not originate within the triad but rather have their source in harsh conditions, external to the group, which burden them all: "These are poor people" (Druze). Here the differences between the older man and the younger individuals are less important than their overriding, equalizing poverty. There is no longer a hierarchy of power within the triad; rather, all group members are similarly powerless in the face of dominating, wounding external circumstances.

In the PM_2 stories, while the age division between the older man and the younger group might be noted, it has no consequences for the ensuing behavior, which is muted and egalitarian in character, as between peers. The older man and the younger group exchange salutations, or they discuss some impersonal topic, such as their "work." A few of these stories concern the meeting between parties of strangers who exchange information about their origins.

Finally, under the PM_4 subheading, minimal but accurate descriptions are given. The basic stimulus features are reported without elaboration. The older man holds his arm out to two younger individuals, though no reason is given to explain either the gesture or the three-person gathering in which the behavior occurs.

Magical Mastery. In these responses, the major card features are misperceived so that the older man's authority is either overstated or blatantly denied.

The first subcategory (MM$_1$) includes all those responses in which, via distortion of major stimulus features, the potential aggression of the older man or of the group as a whole is unreasonably magnified. The older man is viewed as either a murderer or as a victim of some deadly agency. For some respondents, the older man is seen as pointing a pistol at the younger figures, perhaps in retaliation for some theft or in trying to sexually assault a woman (the taller figure of the younger dyad). Conversely, other respondents might see him as completely impotent rather than dangerous, and even on the point of death: He is sick, crippled, or blind. Common, then, to all the MM$_1$ responses is the idea that some deadly force is out of control. The older man is either the agent of destruction or, conversely, the victim of some lethal, disabling process.

In MM$_2$, the potentials for harmony or pleasure are arbitrarily inflated, in the service of denial. The older man dances or puts on a show for the younger group; he reaches out to embrace them; in some cases, he is a divine figure, a god who blesses the younger men, who pray to him.

The Male Authority Card: Intercultural Comparisons

For this card it was hypothesized that the responses from younger subjects, regardless of culture, would fall in those categories that reflected more authoritative, disciplining, and directive stances on the part of the older male figure. Hence we expected that younger men would favor the Active Mastery or Qualified Active Mastery categories over the more passive and magical possibilities. By contrast, we expected that older subjects would favor the more passive versions of the paternal figure and would turn him into a consultant, a peer, or even a supplicant of the younger figures. Authority is a risky business; even the principled ruler is often hated. Accordingly, we expected that older men, needing the assurance of love, would overlook the more flagrant versions of authority and would instead play up the more peaceable, egalitarian, and affiliative card possibilities.

Table A-4 gives the distribution of responses to the Male Authority card by age, by the four cultures in which versions of the card were employed, and by the mastery types. Note that while this distribution is not statistically significant (perhaps because of the highly structured nature of the stimulus), the distribution patterns resemble those refracted by the Rope Climber, Heterosexual Conflict, and Horse and Men cards, stimuli that do generate statistically significant distributions. Thus we see the characteristic elevation on Active Mastery and Qualified Active Mastery for younger men, and the characteristic elevation on Passive Mastery (Constricted) for older men (the distributions are displayed in table A–4). As noted earlier, the Druze data for this card, both cross-sectional and longitudinal, amplify these general trends to statistically significant levels.

Druze. The Druze version of the Male Authority card sponsors more drastic age differences than those found among the Navajo, and these more emphatic intercohort differences are reflected in the Druze age/mastery style distribution, which is significant at better than the .05 level. By contrast to the Navajo, in the Druze case the shift away from versions of Active Mastery and toward Passive Mastery is more decisive. Also, in the Druze case, the increase is seen across most subtypes of Passive Mastery: PM_2 rises in later life at about the same rates as PM_4, and this finding suggests that in the Druze culture a common cause is promoting these two versions of passivity. Thus depression (rather than defenses against aggression) may be projected by elderly Druze in their use of PM_2. The most noteworthy Druze findings concern the contents of the responses coded under PM_4. This index rises in later life because of the rather sudden appearance of stories in which the older man figure is presented as a beggar. Given their intense and stubborn mistrust of dependency, there is no such role for older men in Druze society. Accordingly, this derogatory view of the older man figure by senior Druze cannot refer to cultural sponsorship and must be shaped by personal rather than by social promptings. By captioning the older man figure as a beggar, senior Druze men may distance themselves from dependent wishes that they find unacceptable. In effect, the Druze may use the TAT to live out, in fantasy, those closeted tendencies that move toward psychic prominence in their later life but that cannot be directly expressed in behavior, except in the projective niche provided by the religious subculture. Within the particular extrasocial niche provided by their secret religion, the elderly Druze can be beggars and petitioners of Allah, and these senior Druze may use the opportunity provided by the TAT card to display the kind of need that they bring to Allah—

and only to Allah. However, while there is a rise in PM_4, the fact is that the older Druze, like the older Navajo, keep some title to all the mastery positions. They reorder their mastery priorities without finally abandoning any form of mastery. This catholicity in regard to life-styles is in keeping with Druze circumstances. The older Druze tends to stay active in all major areas of life—he works his fields (though now with his sons' help), he is active in clan, village, and Druze community affairs, and he lives out his needy, dependent side in the special, boundaried precincts of the *hilweh*, the prayer house. Because their life provides them with a god to adore and with sons who defer to them, the elder Druze can live out both the active and the passive sides of their nature, as reflected by this card, without having to make a final choice among these potentially conflicting needs.

Navajo. While the age/mastery distribution of Navajo responses to the Male Authority card does not reach statistical significance, the profile developed by this distribution is consistent with the overall pattern of findings and at the same time presents a particularly Navajo version of them. On the one hand, the interage comparison shows the typical rightward drift in the data, toward Passive and Magical Mastery, with advancing years. On the other hand, the major shift is not from Active to Passive Mastery; emphasis remains within the extended Active Mastery range, with AM_1 as the high point for the younger group, and QAM_1 as the peak for the older group. In other words, the younger Navajo view the older man as a rather harsh and secular authority who asserts his will in the face of opposition. The older men, while they still see him as an authority, imply that his legitimacy has a moral base; his status is based on nurturance rather than coercion.

The Navajo results may be summarized as follows: Older men can retain some authority without undue anxiety and opposition if they move away from direct confrontations with the young over mundane affairs and move into the position of benevolent advisers. Thus the Navajo stories grouped under QAM_1—the mastery location most favored by older Navajo subjects—typically feature an older man who advises younger men to follow the "good path," "the right way," "the beautiful (*hojonje*) way." This "elder statesman" authority mode is in conformity with the arrangement of Navajo traditional life. At least in the traditional camps of the Navajo, the older men are the keepers of the oral tradition and Navajo ritual ways, and they interpret these to the young. While this leadership style of older Na-

287

vajo is in line with cultural prescription, it reflects the deeper currents of the life cycle as well, in that it is more selfless, more tender—more passive, if you will—than that favored by the younger men.

Lowland Maya. This is the one group whose age/mastery profiles run directly counter to our hypotheses for the Male Authority card. Eighty-five percent of the younger men's responses fit the formalism, distance, and restraint called for by the PM_2 category. It is only for the elder respondents that this univocal emphasis on one response style breaks down and the stories distribute into a wider range of mastery styles. It is as though the perhaps fearful control exercised by the younger men begins to break down in later life, and we see a proliferation of all the submerged response possibilities, predicted as well as unpredicted, Passive Mastery and Qualified Active Mastery. This "liberation" from the bondage of anxious restriction seems to occur in the midfifties, and from then on the age profiles do shape according to our predictions. That is, the differences in age/mastery profiles that we expect to find between the 40-year-old and the 60-year-old men become evident when we compare the profiles from the 55- to 69-year-old men against those contributed by the 70-year-olds. Where the 55-and-over group shows some Active Mastery, the 70-and-over group shows none, and the same age difference is noted in regard to Qualified Active Mastery. Meanwhile, the oldest men made a decisive shift in favor of sensual passivity (PM_4) and toward euphoric denial (MM_2). Thus, while the men 55 and over give primacy to MM_1, the paranoid phrasing of Magical Mastery, the oldest men give equal play to the more benign MM_2. In effect, then, as the Lowland Maya age, we see a gradual reduction of the anxiety and control indicators represented by PM_2 and MM_1, with the final emergence among the oldest group of a rather relaxed and affiliative relational style that mingles passive receptivity and self-indulgent optimism. Again, the profile sketched by the oldest men does correspond to the psychosocial ordering of their lives. The elderly Lowland Maya have little or no ritual or political functions; they work as much as their strength will allow; and they are in the main supported by their sons and by other close family members from whom they receive both love and respect. However, this correspondence between TAT profile and actual lifestyle is not as true for the youngest men as it is for the oldest men. The youngest men's profile would lead us to predict their commitments—general in the Mayan culture—toward politeness and inner control, but it does not pick up the counterbalancing energy and zest for work that is the hallmark of the younger Yucatan man.

Highland Maya. The Highland Maya tend to reverse the pattern that the Lowland Maya produce in response to the Male Authority card. While the younger men are bimodal between Active Mastery and Passive Mastery$_2$, the older men concentrate, in a manner reminiscent of the younger Lowland Maya subjects, almost exclusively on Passive Mastery$_2$. Seventy-five percent of the Highland Maya men aged 70 and older favor this particular mastery orientation. Furthermore, responses that fit the criteria for PM$_4$ and MM$_1$ have appeared in the older group, although these were completely lacking among the younger men. The Highland Maya profiles correspond most closely to the form predicted from our original hypotheses, and a two-way chi square, which compares active to passive responses in the below-60 and above-60 groups, is statistically significant at better than the .05 level.

The Highland Maya data fit the ideal, "species" pattern, but they are also consistent with the life of the older Highland Maya peasant. The older man's profile is centered on fear and on passive avoidance of the occasions for anxiety; so, too, is the ordinary life of the elderly Chiapas *campesino.* At any age, the life of men in the Highland Maya pueblo is rounded by hard work, heavy drinking, drunken violence, the struggle for prestige in the *cargo* system, and fear of the envy of others. The heavy, brooding world of the Highland Maya is a risky place for an old man who has lost the appetite and strength for work and physical combat, but who is at the same time feared and envied for the magical powers that he is supposed to possess. As in the TAT, anxious constriction is the keynote of the old Maya's lives. They sit apart, and they take care not to lose control or to offend the envious ones. Thus, while all men seem to move in later life to one or another format of passivity, it is only the fortunate old men—for example, of the Druze and the Lowland Maya—who can also afford the luxury of relaxed passive dependency (PM$_4$). In the less orderly communities, such as those of the Highland Maya, they must express their passivity by "staying quiet," rather than by asking for help.

The Authority Card: Longitudinal Studies

This card was readministered, after a lapse of five years, to forty-seven men of the original Galilean Druze sample. The mastery profiles developed from the Time$_1$ to the Time$_2$ data are shown in table A–5. This display shows age shifts, consistent with a developmental hypothesis, in the rightward direction: Where the Time$_1$ peak stood at

TABLE A–5

The Male Authority Card: Comparison of $Time_1$ and $Time_2$ Responses Given by the Same Druze Subjects[a]

Subjects	Active Mastery	Qualified Active Mastery	Bimodal Mastery	Passive Mastery (Sensual)	Passive Mastery (Constricted)	Magical Mastery	Total
Druze: $Time_1$	34 (.36)[b]	20 (.21)	12 (.13)	7 (.07)	17 (.18)	4 (.04)	94
Druze: $Time_2$	14 (.15)	34 (.36)	15 (.16)	3 (.03)	23 (.24)	5 (.05)	94

[a] This card was not readministered to Navajo informants at $Time_2$.
[b] When the six mastery categories are collapsed to four—Active, Bimodal, Passive, and Magical Mastery—the probability that the resulting distribution occurred by chance is less than .01.

AM_2, at $Time_2$ the profile is bimodal for AM_3 and PM_4. When we examine the changes between $Time_1$ and $Time_2$ within the sample age groups, we find that younger respondents produce the shift from AM_2 to AM_3, while older respondents (those above age 55) are mainly responsible for the shift away from Qualified Active Mastery to Passive $Mastery_4$. In other words, younger men shift away from the unambiguous display of authority registered in AM_2 toward doubt and misgivings concerning the destructive consequences, for themselves and others, of their authoritative qualities. Meanwhile, the older men slip from the position of the nurturant provider subsumed under Qualified Active Mastery, to the position of beggar. Overall then, in regard to their paternal authority, Druze men follow this sequence as they age: They delight in the exercise of authority; desiring affiliation, they begin to question their own authority; they select authoritative roles that, because they imply giving, are rewarded with love rather than hatred; and finally, they eschew authority completely and ask that others take the authoritative, provident role toward them. In this regard, it is particularly significant that nine Druze men who saw the older male figure wearing one or another face of authority at $Time_1$ see him as a beggar by $Time_2$, and that eight of these men who have so drastically revised their perception in five short years are over 60 years of age. As was noted earlier, such a pronounced shift, particularly in the Druze culture, can have only a developmental rather than a secular meaning. In effect, these longitudinal data point to the developmental meanings not only of the Druze data, but of all the transcultural data elicited by the Male Authority card: In three out of the four subject cultures, the respondents reveal age changes in their subjective relationships to their own authority that are consistent with the developmental hypothesis generally and with the postparental role possibilities in their own cultures. The Navajo become benign counselors; the Druze remain as moral authorities to their friends, and in their more private selves they become the children of Allah; and the Highland Maya experience a flight from paternal authority toward a passive position in which security is gained through restriction and immobility rather than humility and gentleness. Men move toward different roles and statuses in later life, depending upon the opportunities and sponsorship provided by their respective cultures; what is predictable—and developmental—is the move away from the role and statuses that require the exercise of aggressive authority.

Summary: General Issues in Male Aggression

Note that the Male Authority card, as well as the others that we have already reviewed, continues to discriminate the subjects, in a predictable fashion, by culture as well as by age. Again, taking a trans-cultural sweep through the data, we find that the cultural variations to some degree replicate the effects of aging, in that they drive all profiles, of the younger and older men alike, in the more passive and receptive direction. Thus the older men of the Navajo trace a profile that, in most respects (save for a lower frequency of Active Mastery$_1$), is a close match to that traced by the younger men of the Druze. By the same token, the older men of the Druze map profiles that suggest more psychic vigor than that displayed by the younger men of either Mayan group and that are decidedly superior to the younger men of the Lowland Maya in this regard. The rankings—Navajo, then Druze, then Maya—are precisely in accord with the rankings established by cards previously discussed, except for the reversal observed between the Lowland and Highland Maya: The Rope Climber and Horse and Men cards ranked the Lowland over the Highland Maya in terms of the prevalence of the more Active Mastery modes, a sequence that does not hold for the Male Authority card. We see again that, as Active Mastery subsides in the profile of both younger and older men across cultures, it is replaced by its frozen, fearful opposite—Passive Mastery$_2$—again confirming our sense that PM$_2$ reflects the culture's as well as the individual's commitment to inhibition of aggression and is not the signature of a "natural," spontaneous current within the self.

Thus far we have been considering, via the medium of projective imagery, our subjects' apperception of the various faces of power: aggression-destructiveness, authority-dominance, energy-impulsivity. And we have found, across cultures, that the readiness to entertain ideas of direct aggression and competition—as well as the possibly destructive consequences of such dangerous urges, for self and others—most clearly distinguishes the younger from the older men, regardless of culture. Thus younger men achieve higher frequencies of stories centered around the physical and verbal contention (AM$_1$) in eight out of a possible fourteen instances. It is only among the Highland and Lowland Maya that older and younger men are tied on this combat issue. In no case does an older cohort provide more stories of the AM$_1$ variety than do younger men. By the same token, younger men lead older men overwhelmingly in stories that have to do with the dangerous consequences of contests and opposition. Younger men have higher percentages than do older men for such stories in the computations for ten cards; they tie the old men on three such computations and take second place to them in only one instance.

The picture seems starkly clear. Across the various and contrasting human communities from which we have gathered data, younger men appear to tell us—at least in their fantasies—that their lives are filled with occasions of challenge, rivalry, and even combat. Then they also tell us that they are moved to accept the challenge, to throw down the gauntlet, to seek out enemies, despite any price that they might have to pay in injury to themselves or guilt over the injury that they might cause others. As we have seen, the older cohorts are not lacking in the more modulated forms of aggression of the sort that can be contained, neutralized within disciplined production and moral leadership; but the younger men are marked and perhaps driven by a particularly reckless sensibility, of the sort that has as its negative counterpart their high profiles on PM_2, the attitude of frozen fear. They do not fear the slings and arrows of the world as much as they fear their own wish to "go in harm's way."

The Denial-eliciting Cards

Presented below are the mastery criteria developed to accommodate the responses to the Desert Scene and Bats and Man TAT cards (figs. 5.10 and 5.11).

The Desert Scene Card

Active Mastery$_1$. The respondent recognizes the stimulus demand, of opposition between human purpose and uncaring nature, and he recognizes that the stimulus represents an arena in which this drama has been worked out. People have tried, in the face of an ungiving environment, to run cattle, to plant crops, and to halt erosion (the hint of a check dam is sometimes seen at the bottom of the gully). People have been active, though not necessarily effective or successful, in the face of a harsh environment.

Active Mastery$_2$. The unsympathetic and harsh nature of the environment is recognized, but compensating human activity is not mentioned. Although human agency is not featured in these stories, they are nonetheless judged to be active responses. The respondent has at least faced, without recourse to denial, the essential desolation of the scene.

Qualified Active Mastery₁. Men are responsible for the blighted landscape: Their attempts at cultivation or overgrazing by their sheep have caused erosion and have produced a desert. Human agency, human assertion, is worse than futile; it brings about the destruction of the environment.

Passive Mastery₁. The barren nature of the scene is recognized as such, but some details are misinterpreted as a sign of oncoming, unchecked disaster. The background cloud may be the smoke from a forest fire.

Passive Mastery₂. Animals submit passively to a vaguely perceived natural disaster. The cow skull may represent an animal "drowning in water."

Passive Mastery₃. Only one or two peripheral details are accurately perceived and named. The rest are ignored, and there is no reference to the total scene.

Passive Mastery₄. The reality of the dry, ungiving desert is denied. Rain is teeming from the background clouds; the gully is a stream bed full of water, perhaps emptying into a lake; the desert expanse is itself a lake or ocean; or the entire scene is one of lush plantation and burgeoning crops.

Magical Mastery₁. The landscape is a setting for uncontrolled natural or human forces: erupting volcanoes, floods that sink ships, sea battles, atomic bomb explosions (the background cloud), etc.

Magical Mastery₂. The card calls up images of comfort and bounty. Fence posts are angels; the cacti are Christmas trees; the gully is a church spire; the cloud is a sack of corn or money; etc.

The distribution of responses by mastery style, age, and culture are summarized in table A–6. The hypothesis, that willful, egocentric denial will replace realistic perception as aging proceeds, is amply borne out. Active and Qualified Active Mastery responses that honor the reality principle are predominant for only the youngest respondents, those in their thirties and forties. For the midlife men, those in their fifties, there is a decisive shift away from Active Mastery toward Qualified Active Mastery, and toward the primacy of defensive fantasy, as represented in Passive Mastery₄. By the seventh decade, Active Mastery responses have almost completely disappeared, and

FIGURE A-5
The Desert Scene Card: Distribution of Responses
by Age and Mastery Style (Navajo, Druze, and Mayan Subjects)

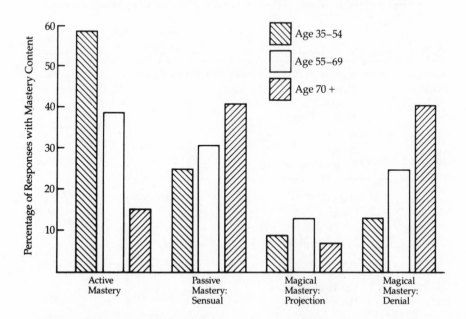

the highest response percentages are now found in Qualified Active Mastery, and "fantasied nurturance" (PM$_4$ and MM$_2$) categories: Rain redeems the desert. For the seventies-and-over group, the responses cluster almost exclusively in the categories of fantasied nurturance: Wish alone makes the desert bloom. Table A-6 presents the actual age/culture/theme frequencies developed by the Desert Scene card, and figure A-5 depicts the distribution by age and theme in more graphic form.

Clearly, with each succeeding decade after the forties, the ambiguous but troubling stimulus is increasingly distorted in a wishful, self-comforting manner.

The Bats and Man Card

Active Mastery,. The threat from malevolent creatures is recognized, and the hero takes some active, possibly effective, actions against the attack (Navajo: "He is pretending to be asleep; when those birds get close, he'll reach up and grab one").

TABLE A–6

The Desert Scene Card: Distribution of Responses by Mastery Style, Age, and Culture

Mastery Style	Subjects	Age 35–54		55–69		70+		Total
Active	Navajo	9	24 (.17)*	2	3 (.03)	0	2 (.04)	
Mastery	Druze	13		1		2		
(AM₁, AM₂,	Lowland Maya	0		0		0		
AM₃)	Highland Maya	2		0		0		
Qualified	Navajo	20	57 (.41)	18	38 (.34)	1	5 (.11)	
Active	Druze	21		13		3		
Mastery	Lowland Maya	10		3		0		
(QAM₁, QAM₂)	Highland Maya	6		4		1		
Passive	Navajo	3	16 (.12)	3	7 (.06)	1	3 (.07)	
Mastery:	Druze	6		3		1		
Constricted	Lowland Maya	6		1		0		
(PM₁, PM₂, PM₃)	Highland Maya	1		0		1		
Passive	Navajo	6	17 (.12)	13	26 (.23)	5	15 (.33)	
Mastery:	Druze	1		2		3		
Sensual	Lowland Maya	6		11		5		
(PM₄)	Highland Maya	4		0		2		
Magical	Navajo	2	9 (.07)	3	12 (.11)	0	3 (.06)	
Mastery:	Druze	4		6		1		
Projection	Lowland Maya	2		3		2		
(MM₁)	Highland Maya	1		0		0		
Magical	Navajo	1	15 (.11)	4	26 (.23)	4	18 (.39)	
Mastery:	Druze	7		14		8		
Denial	Lowland Maya	2		4		3		
(MM₂)	Highland Maya	5		4		3		
TOTALS			138		112		46	296

* The chi-square test was applied to the cell totals. The probability that this distribution occurred by chance is less than .001.

Active Mastery₂. The threat from malevolent creatures is recognized and elaborated; however, the hero does not take any effective action against the obvious danger.

Qualfied Active Mastery₁. The hero, by some aggressive or immoral act of his own, has stirred up a flock of evil creatures who will destroy him.

The Denial-eliciting Cards

Qualified Active Mastery₂. The recumbent man is not recognized as such; instead, the respondent describes a flock of birds and their actions. However, after the human figure is pointed out to him, the respondent goes on, spontaneously, to tell a story in which a recumbent man is attacked by a swarm of malign creatures.

Qualified Active Mastery₃. The man and the birds are accurately perceived, but the birds are construed to be figures in the hero's dream. In stories of this sort, the hero may awake to find that he was "attacked" by his own bad dream rather than by real creatures.

Passive Mastery₂. The human figure is overlooked, and the birds are seen in rather aimless flight ("just flying around"). When the figure of the recumbent man is pointed out, the subject cannot or will not see it, and does not go on to relate a story in which a man is attacked by harmful creatures.

Passive Mastery₃. The human figure is not recognized, even after it is pointed out to the respondent; instead, the birds are seen to be engaged in benign, center-seeking, comfort-seeking activities (they are returning to their nest; they are bringing food to their nestlings; or they are looking for fruit, seen growing on the bushes). Instead of being seen as predators, the birds are seen as relatively benign, needful, even vulnerable creatures.

Passive Mastery₄. The subject recognizes, accurately, one or two minor details, commenting on these without interpreting the card as a whole.

Magical Mastery₁. The whole scene is dynamic and charged with aggression. The birds are demonic; they fight among themselves. The man is deformed. The "bushes" are puffs of smoke from explosions. In some cases, the birds represent the judgment of God on an evil person.

Magical Mastery₂. The man is recognized (either spontaneously or after being pointed out), but no conflict is depicted between him and the birds. Instead, they are protective figures, for example, angels who come to bless or to feed him. In other cases, even after being pointed out, the man is not noted, and the birds are seen as vulnerable, pleasing, and protective figures—butterflies, "pretty little birds," or cherubs.

In sum, applied to the data of this card, the mastery types define a clear continuum, ranging from stories in which the hero "takes arms against a sea of troubles," to those in which the very idea of

FIGURE A–6

The Bats and Man Card: Distribution of Responses
by Age and Mastery Style (Navajo, Druze, and Mayan Subjects)

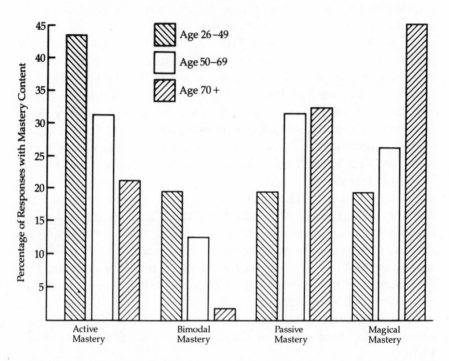

trouble is strongly denied, replaced by themes of love, harmony, and interdependence. As with the Desert Scene card, we find that younger men, in the age range 26–49, give, by a ratio of more than 2 to 1, the highest priority to realistic perceptions; by the fifties, the balance between realistic and wish-fulfilling card perceptions approaches equality, though with the weight still in favor of realism; by the sixties, the balance has shifted decisively in favor of determinedly benign but unrealistic responses; and by the seventies, more than one-third of all responses fall into the Passive Mastery category, focused on harmless, feeding, domesticated birds. An illusory comfort has been gained, but at the expense of realistic evaluation (see fig. A–6).

Tables to Chapter 5

The following tables (A–7 through A–12) pertain to chapter 5, "The Season of the Senses." They present, in tabular form, the various distributions that are shown in graphic form in chapter 5. They are presented in this form for those readers who wish to inspect the detailed age/culture numerical distributions.

TABLE A–7

Sources of Contentment: Distribution by Age and Culture

Age	Culture	Active Productivity		Active Pleasure		Domestic Sentience		Passive Receptivity		Totals
35–54	Druze	23		4		25		14		
	Navajo	8	41 (.53)[a]	4	9 (.20)	18	45 (.46)	9	26 (.30)	121
	Highland Maya	10		1		2		3		
55–69	Druze	12		6		15		22		
	Navajo	12	26 (.34)	13	21 (.47)	9	26 (.27)	12	36 (.40)	109
	Highland Maya	2		2		2		2		
70+	Druze	4		4		16		21		
	Navajo	3	10 (.13)	8	15 (.33)	8	26 (.27)	3	26 (.30)	77
	Highland Maya	3		3		2		2		
TOTALS		77		45		97		88		307

[a] The chi-square test was applied to the cell totals. The probability that this distribution occurred by chance is less than .001.

TABLE A–8

Distribution of Navajo Dreams
by Mastery Style and Syntonic Orality Scores

Mastery Style of Dream	Subject's SOS Below Median	Subject's SOS Above Median	Total
Active Mastery	17 (.68)[a]	8 (.30)	
Passive or Magical Mastery	8 (.32)	19 (.70)	
TOTALS	25	27	52

[a] The chi-square test was applied to the cell totals. The probability that this distribution occurred by chance is less than .01.

TABLE A–9
Median Distribution of Navajo Syntonic Orality Scores
by Health Status and Age

Syntonic Orality Score	Age	Health Status I (Healthy)		Health Status II (Moderate Chronic Illness)		Health Status III (Potentially Terminal Illness)		Total
SOS equal to or less than .18 (median)[a]	35–54	10	24 (.69)[b]	8	25 (.57)	2	8 (.24)	57
	55–69	12		10		3		
	70+	2		7		3		
SOS equal to or above .19 (median)	35–54	2	11 (.31)	4	19 (.43)	8	26 (.76)	56
	55–69	3		8		10		
	70+	6		7		8		
TOTAL		35		44		34		113

[a] SOS estimated from $Time_1$ data; health status estimated at $Time_2$ from United States Public Health Service clinical records.
[b] The chi-square test was applied to the cell totals. The probability that this distribution occurred by chance is less than .001.

TABLE A–10
Responses to Interview Question "Who was the most
important person when you were a child?":
Distribution by Age and Culture

Age	Culture	Fathers or Older Male Kin		Mothers or Older Female Kin		Total
30–59	Kansas City	12	21 (.70)[a]	20	21 (.48)	
	Highland Maya	9		1		
60+	Kansas City	4	9 (.30)	19	23 (.52)	
	Highland Maya	5		4		
TOTALS			30		44	74

[a] The chi-square test was applied to the cell totals. The probability that this distribution occurred by chance is less than .06.

TABLE A–11

The Boy on Cliff Card: Distribution of Responses
by Age, Theme, and Culture

		Total			
Theme	Culture	30–54		55–82	
Hero looks out with	Navajo	25	28 (.67)[a]	16	18 (.40)
productive or combative	Lowland Maya	2		2	
intent: for new lands,	Highland Maya	1		0	
better grazing, lost					
livestock, or enemy					
Hero has come to high	Navajo	7	14 (.33)	19	27 (.60)
point to see beautiful	Lowland Maya	5		5	
country	Highland Maya	2		3	
TOTALS			42		45

[a] The chi-square test was applied to the cell totals. The probability that this distribution occurred by chance is less than .05.

TABLE A–12

Age Distribution of All TAT Responses with Visual Content[a]

	Age			
Responses	35–54	55–69	70+	Total
Implied visual activity: hero looked at, hero looks at threat, etc.	54 (.18)[b]	48 (.21)	19 (.16)	
Functional looking: in service of production, competition, or mobility	107 (.37)	57 (.26)	26 (.20)	
Nonfunctional looking: hero looks, without function or feeling	108 (.37)	88 (.40)	65 (.52)	
Pleasurable looking: hero looks at pleasing sights	22 (.08)	28 (.13)	15 (.12)	
	291	221	125	637

[a] All responses with references to visual activity given to all TAT cards.
[b] The chi-square test was applied to the cell totals. The probability that this distribution occurred by chance is less than .005.

The following tables (A–13 through A–16) pertain to chapter 6, "The Inner Liberation of the Older Woman: Psychological and Fantasy Measures." They present, in tabular form, the various distributions that are shown in graphic form in chapter 6. They are presented in this form for those readers who wish to inspect the detailed age/culture numerical distributions.

TABLE A–13
The Family Card: Distributions of Role Descriptions
of the Older Woman Figure by Age, Sex, and Social Class

Category	Social Class	Age 40–54		Age 55–70	
		Men	Women	Men	Women
Submissive,	Middle class	10	6	1	3
nurturant	Working class	4	4	2	3
	TOTAL	24[a]		9	
Controlled by	Middle class	1	9	1	0
older man	Working class	5	5	1	2
	TOTAL	20		4	
Limited by	Middle class	2	3	1	1
children	Working class	3	2	2	4
	TOTAL	10		8	
The good	Middle class	2	0	7	2
mother	Working class	7	0	3	2
	TOTAL	9		14	
The matriarch	Middle class	1	3	0	3
	Working class	1	0	1	1
	TOTAL	5		5	
Hostile,	Middle class	2	1	4	3
self-assertive	Working class	1	1	5	4
	TOTAL	5		16	

[a] The chi-square test was applied to the category totals. The probability that the distribution occurred by chance is less than .001.

TABLE A–14
The Family Card: Distributions of Role Descriptions
of the Older Man Figure by Age, Sex, and Social Class

Category	Social Class	Age 40–54		Age 55–70	
		Men	Women	Men	Women
Altruistic	Middle class	5	9	2	1
authority	Working class	3	3	1	3
	TOTAL	20[a]		7	
Assertive,	Middle class	2	0	2	2
but guilty	Working class	6	1	1	1
	TOTAL	9		6	
Formal	Middle class	5	5	0	0
authority	Working class	2	3	1	0
	TOTAL	15		1	
Surrendered	Middle class	2	3	4	3
authority	Working class	6	1	3	2
	TOTAL	12		12	
Passive,	Middle class	0	4	2	6
affiliative	Working class	3	4	3	8
	TOTAL	11		19	
Passive,	Middle class	4	1	4	1
cerebral	Working class	1	0	6	2
	TOTAL	6		13	

[a] The chi-square test was applied to the category totals. The probability that the distribution occurred by chance is less than .001.

TABLE A–15
Mastery Styles for Kansas City
Females by Age

Mastery Styles	Age (N = 144)		
	40–49	50–59	60–70
Active Mastery			
Rebellious daughters	11	16	7
Moralistic matriarchs	6	8	4
TOTAL	17[a]	24	11
Passive Mastery			
Maternal altruists	2	7	2
Passive aggressors	5	13	13
TOTAL	7	20	15
Magical Mastery			
TOTAL	8	16	26

[a] By the chi-square test, the distribution of Active, Passive, and Magical Mastery totals by age groups is significantly different from chance at the .02 level.

TABLE A–16
Extrapunitive and Intropunitive Kansas
City Women by Life Satisfaction Ratings

	LSR at or below the median	LSR above the median
Intropunitive	12[a]	3
Extrapunitive	13	23

[a] By the chi-square test (Yate's correction), this distribution is significant at the .02 level.

NOTES

Chapter 1

1. Sampling error, particularly in clinical work, can contribute to a scientific bias against the possibility of psychological development in later life. There is not much of a naturalistic tradition in geropsychology, and consequently the psychologically intact, community-dwelling aged are rarely studied in their natural settings. In addition, fearing that they will be labeled as crazy, the moderately distressed aged do not, particularly at the time of onset, usually bring their problems to psychologists. Instead, while unconsciously hoping for psychological relief, they often translate emotional problems into somatic symptoms and bring these to the attention of a caring internist. Medical personnel tend to collaborate with this somatic defense and will too often concentrate on the symptom without recognizing the underlying anxiety or depression. The incidental concern shown by the internist can help maintain psychological stability in many older patients, but those who break down despite such consideration are referred to the psychiatrist or clinical psychologist only when the disease has reached critical, even irreversible proportions. As a result, the mental health practitioner does not get to see a range of distressed elderly, from the mildly troubled to the wildly psychotic, but instead mainly sees those who are *in extremis*. In effect, mental health practitioners are more apt to treat the terminal patient, in whom all major systems decompensate, rather than the *aging* patient (who is, save for some usually treatable difficulties, in a relatively stable condition). The overrepresentation of terminal patients in clinical practice leads to a skewed sample and to a particularly morbid picture of the older patient in general. In effect, geropsychiatrists and clinical geropsychologists confound the state of termination with the quite distinct state of aging, in such a way that the terminal picture tends to become the exemplar of the aging condition. Thus aging becomes associated, even in normal populations, with profound, accelerating, and irreversible loss.

2. The surgent phase is particularly evident in puberty. As if awakening from a long sleep, the child's body begins the transformation into the configuration of a potential parent; the body acquires adult mass, menarche is passed, secondary sexual characteristics appear, and on the psychological plane the matching aggressive and sexual drives surge toward consciousness. In reaction, boys and girls either become obsessively concerned with each other, or they involve themselves, somewhat frantically, in various "purified," asexual—usually politico-religious—activities.

Chapter 2

1. I have reported the various field studies and their findings separately (Gutmann, 1964, 1968, 1969, 1974).

2. This last expectation adds to the difficulties of interviewing an older *aqil* about his childhood experiences. He might become remote or evasive, and at times even angry: "Why do you ask me about the time when I was ignorant, before I knew God, before I became close to God? I was like an animal then, I did not think of God, and it is a shame to think of such things!"

3. Thus Freud did not only generate new theories of the mind; his great contribution was in the generation of new data—dream images, free associations to dreams, slips of the tongue—that had no prior cognitive status but are now routinely used by his critics to refute Freud and to construct new theories.

Chapter 3

1. Neglecting the time dimension, most studies that utilize both projective and more "objective" self-report tests fail to report correlations among the data developed by these instruments. Jacobowitz's work, being more sensitive to the functions of fantasy as an alternative to overt action, shows that the correlations between fantasy and objective data do exist but are revealed only over time, after fantasy has exercised its proper function of delaying (and rehearsing) eruptive, potentially dangerous behavioral potentials.

2. The sample of artists, though selected opportunistically, is representative of various styles, historical periods, and media. I report on those masters whose collections became available to me in accidental ways, that is, I was able to view a retrospective exhibit, or I chanced upon a published collection of works by an artist who met my criterion for longevity by surviving into at least the seventh decade of life.

Chapter 4

1. It is therefore hard to predict a correspondence between the forms of fantasy and the patterning of overt, socially mediated behavior. Because fantasy measures do not predict to current behavior but to behavior as much as five years in the future, positive correlations between unconscious imagery and action are picked up only in longitudinal studies, over time. Fantasies are, after all, the mental picture cast up by peremptory wishes (or by our fears about them). We can reasonably expect that desires universal in their distribution—the aggressive wishes of young men or the receptive wishes of old men—will eventually find their way into behavior via the age-graded prescriptions that govern the norms and forms of public behavior.

2. If these predictions are confirmed, our projective instrument is incidentally validated. Success would mean that the fantasies picked up by the TAT instrument are more than the artifacts of a particular technique. Instead they register imperious action tendencies that will eventually filter into socialized behavior via their reciprocal cultural norms.

3. Further evidence for the universality of young male violence is found closer to home by Conrad Arensberg and Solon Kimball (1968), ethnographers of Irish peasant life. The relationships between younger and older men discovered in their study are reminiscent of both Samburu and Comanche. In the Irish villages, the young men are voluble in voicing disrespect for the aged, but only when they are out of their presence. Then, they claim to see no difference between the deliberations of the older men and the gossip of old women. But while the younger men are more combative, the older Irish peasant, like the older Samburu, still wields the real social power. These authors

Notes

find that the older men take precedence in formal and informal social gatherings, and receive the largest share of delicacies laid on at such meetings.

These observations are confirmed by Gordon Streib (1968), who finds that the bachelor group of young Irish men, aged 18 to 35, is devoted (when off work) to drinking, dancing, gambling, and brawling. As Streib sees it, such fun and games keep the young men—called boys until they marry—from interfering in prestigious affairs, such as the older men's rule of the community, and continuous emigration of the "wild geese" (the excess, potentially unruly young men) has the same result. A proper sociologist, Streib understands the violent nature of the young as a side effect of the repressive social arrangements in traditional Irish society. But I would argue that the younger male tropism toward violence is universal and primary, and that the solid bloc of elders is organized partly in opposition to the younger men and their appetite for mayhem. They are reacting against the threat, not bringing it about.

4. Further evidence that the age differentials in the distribution of aggression and control are inborn and prior to the social customs that express them, is provided by Keith Lovald (1962), who studied the social life of the Skid Row aged. Even in their relatively anarchic precincts he finds that older men—particularly those over 60—will even cooperate with their traditional enemy, the law.

Statistics on civil and domestic crime also reveal the age distribution of male violence: Young men can be dangerous. Thus Richard Gelles (1973) tells us that men in the 41–49 age group are most likely to be brought to law as wife beaters and that the average age of the beaten wife is 37. By the same token, the jail population is overwhelmingly male (95 percent), disproportionately black (42 percent), and decidedly young—60 percent were under 30 years old.

5. Older men of aggressive, hunting societies are at particular psychological risk when modernization undercuts their ritual routes to compensating supernatural power. The key example comes from John Honigmann (1954), on the Attawapiskat Cree tribe of northern Canada, a group that lives by hunting and trapping, and values strenuous masculine activity. Their physical debility causes severe psychological problems for older Cree men, particularly now that the belief in sorcery, their traditional route for asserting power, has been undercut by the Christian missionaries.

6. Wolff (1959), who served as a physician in China for many years, reports that members of the older generation are financially supported by their children, in accord with the dictates of the Confucian religion. Untroubled by this reliance on their own offspring, the elders seem happy and serene, as their dependent status is sanctioned by the social norms. This happy congruence between the social and personal requirements of the older Chinese has a healthy outcome: In the course of his clinical work, Wolff notes, his chronic brain syndrome patients rarely showed any psychotic reactions. This lack of drastic CBS symptoms is contrary to Western clinical experience, as is the serenity with which the elderly Chinese accept oncoming death. They see it not as a termination but as an entry to a better life. Not only have they turned aside from secular affairs, but they have accepted death, the ultimate passivity, with a pleasure that goes beyond mere resignation.

7. The older leaders, regardless of setting, maintain the more stereotypically traditionalist or folk position, against the more time-bound "industrialized" mentality of the younger *politicos*, a finding that suggests that generational, rather than developmental, factors could account for these differences. However, the comparative design of the study helps to rule out the secular effects. The latter would account for the age differences in values observed among the leaders of Medellin, where cultural change has been rapid, but not among the leaders of Popoyan, where cultural change has been slow, and where both younger and older men have been equally exposed, in the course of socialization, to traditionalist values.

8. Hans Thomae (1962) comes up with matching evidence from the European professional class. Whereas younger German white-collar workers (aged 35 to 49) look for excitement, contact, and open futures in their jobs, older men (aged 50 and over) look for stability, tempered by some guarantee of independence from the boss. Having given up occupational striving and on-the-job confrontations, the older men appear to be more relaxed. Thomae noted that those over 50 have a greater capacity for "positive compromises with reality," as well as less psychosomatic disease than their younger

co-workers. Thomae's findings are confirmed by Bergler (1961) in a West German population, and by Alfred Heron and Sheila Chown (1962) among older English workers: The latter report that British supervisors and foremen prefer older workers, who cooperate with management, over the younger men, who "work for themselves."

9. In Western, secular cultures, such beliefs of divine sponsorship are coded as delusions. In Sigmund Freud's (1911) account, for example, Shreber was brought to treatment because of a belief, cresting after age 50, that God was turning him into a woman. Freud may have thus presented us with the first account on record of an exacerbated midlife crisis, in which the patient holds God accountable for his own later-life "feminization."

Chapter 5

1. Food-centered statements were assigned weights that reflected the intensity of oral need presumably expressed through them. Mentions of food production, marketing, or preparation were given lower scores than mentions of actual eating or drinking. By the same token, mentions of food consumption by others were scored lower than mentions of food consumption by the respondent himself. Each expression of oral interest was also mapped on various formal coordinates, for instance, by time location, depending on whether the oral reference was to the present ("I now eat . . ."), to the respondent's youth ("As a boy, I ate . . ."), to his childhood ("My mother fed me . . ."), or to the historical past of his group ("The people used to eat . . ."). I also distinguished between oral references in the interview proper and those found in the subject's more "projective" utterances: his TAT stories or his dream reports. In addition, I distinguished between positive ("I like to eat") and negative statements: those cases in which foodstuffs or food-related activities were described as disagreeable, harmful, or taboo ("These foods make us ill"). The sum of these weighted references constituted the respondent's total orality score (TOS). The raw TOS was not in itself useful, as a high score could be the artifact of a particularly long interview and not a measure of strong oral need per se. Accordingly our final comparison—between generations, between regions, and between cultures—employed a set of standardized scores, each expressed as a percentage of the TOS.

2. Other accounts from other regions spell out the association between gerontocracy, food taboos, and "gero-orality." Yap (1962), reporting that Chinese patriarchs have greater control over their families, also notes that they receive large portions of the ceremonial feasts. Charlotte Ikels (1975) agrees, stating that "becoming a village elder (in traditional China) entitled one not only to an extra share of meat at all ritual feasts, but also to the unqualified respect of one's juniors." The association between feasting and prestige reaches far beyond the Orient: Prins (1953) reports that the most exalted grade of Kikuyu (East Africa) elders had the sole responsibility for dealing with breaches of taboo and that they alone could eat of the sacrificial offerings. Likewise, social power won first place at the table among the Bering Eskimo: Lantis (1953) reveals that the Nunivak elders were so revered that younger members of the tribe could not speak their names, and distribution of the choicest foods proceeded according to a ladder of age, starting with the eldest. According to Alan Holmberg (1961), the Siriono elderly, because they have survived the rigors of life in the Amazonian Basin, are believed to have special powers: They are not to be insulted, and they are always served first, with the choicest morsels. Similarly for the Navajo: Jerrold Levy (1967) tells us that the aged are most likely to have the special knowledge that is central to the singer's role, and that in former times they ate special and preferred foods that were supposedly beneficial to them but dangerous to the young.

3. The Druze early memory data have not been analyzed to test for maternal preeminence with increasing age.

4. Responses to Rorschach images, analyzed in terms of age, tend to replicate this trend. As compared to younger men, older Navajo produce a notably higher percentage of "water" images: oceans, swimming animals, plants growing in and reflected in water. When the age distribution of respondents above and below the median "water response"

percentage is statistically tested, the age differences are statistically significant beyond the chance level. This tendency to discover water and objects immersed in water suggests that older respondents may be symbolizing a deeply unconscious wish for a return to their intrauterine beginnings, as part of the mother.

5. We should also note that the profound shift in content toward more "regressive" and oral themes in the work of Price and Steig is purely a psychological event and does not reflect an equivalent regression in brain functions. Certainly there is no evidence of disability in the later work of these two artists. Thus, while Price's interests change over the years, his magnificent draftsmanship and characteristic style do not. And Steig becomes, if anything, a better artist in the later years: His images are less conventional and his line is more sensitive, personal, and evocative. Clearly, the degree of creativity is at least conserved in the case of Price and even increases in the case of Steig; what changes most drastically in the later years are the psychological concerns that call for creative management by the aging artist.

6. The intrinsic convergence between eating and looking, proposed by psychoanalytic theorists, directly influences the uses of leisure by the aged. According to Katherine Clancy (1975), those elderly who are most addicted to television-watching are also heavily addicted, in ways potentially injurious to their health, to snack or "junk" food.

Chapter 6

1. This finding sheds some light on the origins of the idiosyncratic, Magical Mastery response. Some critics would argue that such responses reflect organic brain syndromes (particularly in older patients), rather than more dynamic, motivated, reversible processes. If so, then such eccentric responses cannot be interpreted as a signature of changing psychological processes in later life. However, since there is no reason to believe that older Druze women are any more prone than their mates to suffer brain damage, their special predisposition toward idiosyncratic thinking probably has functional rather than organic causes. If so, it is likely that most Magical Mastery responses, including those given by men, have a similarly functional character and can be analyzed, as sub-rosa communications, for their psychological meaning.

Chapter 7

1. The ascendant qualities found among senior American women are not limited to the WASP population. Interethnic comparisons reveal that, regardless of subcultural variations in sex-role standards, the older ethnic woman usually takes on the feisty qualities that were previously denied to her. Despite the traditional machismo of Latino culture, Olen Leonard (1967) finds that the older *chicano* woman is very important in the home. She may defer to males in public, but "the opinions and judgments of the homemaker in decisions affecting family affairs become increasingly important with age." This crucial advisory and judgmental role of the mother continues long after the daughters establish their own families, away from the parental homestead. (Leonard finds an equivalent "implicit matriarchy" among the elder women of rural Mexico.)

Just as "feistiness" is found outside of the WASP majority in North America, it is also found in other parts of the Western world. Robert Havighurst (1960), for example, found that in Germany female interest in local and national politics increases with age, whereas male interest in the affairs of state remains, at best, constant over time. Again, this can hardly be a secular effect, as the older females were raised at a time when the education of women did not prepare them for political life or political interests.

The older woman's vital spark is well captured in Peter Townsend's (1957) description of aged English folk on an outing. The women, he writes, "clearly had more bounce." More objective findings suggest that, for women, "bounce" and longevity may go hand in hand. Robert Kleemeier (1960) states that elderly women in the United Kingdom, institutionalized in mental hospitals, die less frequently, despite the higher female admission rate, than their male age mates.

2. The power of the matriarchal mother-in-law is not limited to her daughter-in-law. She can tyrannize over the son-in-law as well. Among the Iroquois the son-in-law was expected to please his mother-in-law by providing meat for the wife's family. If the mother-in-law was not satisfied with his provision, she could bring about an annulment and banish him from the communal longhouse.

3. Social change may bring about this sort of hectic and extreme response to the aggressive older woman. Returning to Manus after twenty-five years, Mead found:

> What was once a highly specific kinship function in which the paternal aunt and her descendents had special ceremonial powers, based on a special set of ghosts, over the descendents of her brother's children, involving both blessing and cursing, has disintegrated into what might easily develop into a general fear of old women, only a step away from witchcraft.

4. The "phallicism" of older women may also be manifested in their psychoanalyses. Ralph Kaufman (1940), a practicing analyst, notes a renewal of clitoral masturbation in postmenopausal women. He speculates that this development may signal a devaluation of the female genitals and a renewal of fantasies around the possession of a penis.

5. According to Fortes, the old women of the Tallensi are not much better while still alive. He finds that filial obedience toward them is based on mingled respect and fear: One wishes their blessing but fears their curse. Moreover, "old women become completely egoistic; all they think of is having enough to eat and being comfortable, and if they are crossed they are apt to curse anybody incontinently."

6. Certain crucial similarities across the primates allow us to make what are more than arbitrary interspecies comparisons. Most significantly, human and subhuman primates have in common the long dependency and vulnerability of infancy and childhood. In both the ape and human cases, we have begun to discover that the primate infant may have developed communications and social skills that ensure favorable emotional response from caretakers, but it is completely lacking in the skills that ensure physical security: It cannot move in any coherent fashion, it cannot remove itself from danger, and it can barely obtain its mother's milk. Among the lower primates, the infant has, at best, the guaranteed capacity to cling, during the first months of life, to its mother's fur. Of necessity then, adult primates, whether human or otherwise, have in common an intense concern with parenting.

7. While the cross-cultural citations arrayed in this chapter do not finally "prove" our developmental case, neither do they weaken it; and while the confirming evidence does not come from a methodologically clean sample, the sheer weight of the cross-cultural evidence is in our favor—particularly since it bears out a prediction made on the basis of studies originally done in a single culture (a prediction made long before the transcultural evidence cited in this chapter was collated). Thus hypotheses developed from American data alerted us to the possibility of widespread later-life matriarchy, and the review of the cross-cultural data has not been disappointing.

Chapter 8

1. Many piscine and insect species send forth their hordes of untended offspring; in such cases, the statistical probabilities, rather than parental concern, may guarantee the survival of enough offspring to maintain population levels. But human females have a relatively short breeding span and rarely birth more than one child at a time. In our case, individualized parental care becomes particularly important in securing the survival of viable, potentially parental descendants.

2. Responding to the contemporary denigration of parenting, Erik Erikson recently asserted—much like this author—that there is a parental imperative. Despite its association with the sexual drives, the surgent drive to be parental is a potent and independent motive that is now coming under repression in Western society, even as sexual drives are liberated from procreative goals. Just as sexual repression characterized the Victorian era, there is currently a "real danger that a new kind of repression may

become a mark of adult life." Erikson urged psychoanalysts to be alert to the potential harm of such repression.

Indeed, his warning appears to be well-taken. At Northwestern University Medical School, in our studies of late-onset psychiatric disorders, we find that childless older women are represented in the population of first-onset psychotics far beyond their representation in the population as a whole. Replicated in British and Scandinavian samples, these findings suggest that the childless state can have malignant consequences long after the original contra-parental decision was first made. Like any other important motive, parenthood can be denied or repressed but only at one's peril; the subjective will to denial does not negate the objective importance of the denied wish.

3. This folk wisdom is borne out by more sophisticated psychological studies. Thus the experiments of Harlow and Mears (1979) with young chimps raised on terry-cloth "mothers" as well as the observations of René Spitz and Margerie Wolf (1946) on institutionally reared foundling infants demonstrate very clearly that physical security does not, by itself, guarantee the survival of the equally helpless, equally vulnerable human or nonhuman primate infant.

Harlow's baby chimps, for example, had all the food they needed, as well as tactile comfort of a sort. But while they survived to physical maturity without emotional nurture, they could not be parents. Male chimps raised without mothers did not know how to approach females, and similarly raised females would not accept the male for coitus; even when artificially inseminated, they would reject their own offspring. In their case, physical nurture without emotive caring had guaranteed individual but not species survival. In the second case, as reported by Spitz, the foundling infants, tended by efficient but unsentimental caretakers, often developed clinical depressions and were very prone to die of minor children's diseases that normally reared infants routinely shrug off. In their case, what appeared to be adequate physical care did not guarantee even individual survival. Thus feeding and blanketing are not enough; the infant who does not experience a loving welcome into the world probably will not thrive or grow to be the adequate parent of viable children.

In normal child-rearing practice, physical and emotional care imply and complement each other. Adequate physical care can come only from a concerned caretaker; truly adequate emotional care requires a physically secure environment. Love alone cannot comfort a baby who is chronically cold, wet, hungry, or in pain. And the mother cannot do her job, of comforting the child, unless she has, usually in the husband, a reliable provider of physical security.

But while the two forms of security, physical and emotional, are mutually supportive, the primary providers of these must, to a large degree, be distinct from each other. The fossil record now suggests that our species has had as many as four million years in which to evolve our distinctive ways, and particularly our unique parenting practices. To be successful, human parenthood, despite any local variations, had to take account of the unique vulnerability of the human child, and the environmental threats menacing that child—from animal predators; from human enemies; from severe weather; from famine, drought, and plague. Under these stringent conditions, which still hold for a large part of the earth's population, the parent responsible for physical security cannot provide adequate emotional security, and the parent responsible for the provision of emotional security cannot guarantee the logistics of physical security.

The exclusivity of parenting roles is rooted in a fixed, necessary distinction between the placement of the protectors and the protected: Physical security is not guaranteed within the home range, but on its perimeter. Under average-expectable human or primate conditions of marginal food supply and external threat, adequate nourishment cannot be found or even produced within the immediate domestic precincts. Under conditions of foraging or marginal agriculture, local food sources may be quickly exhausted, and providers must fan out, far beyond the home site, to find new game preserves or to open up new lands for herding and cultivation. But even when there is settled agriculture, with some guarantee of adequate food supply, physical security requires that human and natural enemies be kept at bay, and as far from the community core as possible. A community whose defense line is pitched close to its center is, by definition, at chronic risk.

It follows, then, that the physical security of the infant can be guaranteed only by the parent, the father, who is ready to leave home if called on, as hunter, as slash-and-burn agriculturist, or in defense of an outlying borderline. Clearly, those assigned such responsibilities would be violating the basic criteria for physical security if they packed vulnerable children with them on their forays; nor could they adequately perform these vital tasks if they stayed homebound, within the range of a child's cry. Hovering close enough to provide the kind of emotional security on which the child's sense of basic trust is founded, they would inevitably fail to provide the same child with adequate physical security.

The same sorts of iron limitations that dictate that the protected must be separated from the providers of physical security hold also for the provision of emotional security. Whatever else she might be doing, the provider of emotional security should remain within sight and sound of her children or at least be in contact with alternate caretakers who are maintaining that immediate touch. She cannot depart too far from the domestic zone, from those familial settings in which the child's sense of basic trust is generated. Moreover, if the child is to feel relaxed and secure, the mother must feel relatively safe and cherished as well; accordingly, she should carry on her maintenance activities in some relatively protected domestic enclave, one in which her own sense of security can be communicated to her children.

4. I do not mean to disparage women's martial courage or to conserve blood-lust exclusively for men. While women have fought in many wars (usually on a voluntary basis), there is an unwritten though universal law about battle: "When it comes to slaughter, you do not send your daughter." As a consequence, at least a thousand men fall every day, around our planet, in one or another version of armed combat: border scuffles, attempted coups, terrorist raids, tribal wars, duels, fights between criminals and police, as well as fights among rival gangs. Idealistic women sometimes enlist in revolutionary bands, but they are otherwise spared our species' routine work of slaughter. This daily blood tax is levied mainly on young unmarried men, and for a good reason: A large proportion of community males can be sacrificed to warfare and other risky activities before there is a significant fall off in the next generation's replacement rate.

5. These speculations are to some degree confirmed by Martin Hoffman (1970) and his associates. They found that new parents move away from high ideals in favor of amoral familism. The child's needs and potentials are exalted over the more abstract claims of the larger community.

6. We can propose only a tentative explanation of the significant effect that age plays in determining the choice of postparental mastery styles. It may be that late fathers have always been invested in the denial defense—particularly denial of their own aging—and that they maintain such postures in their late middle years through protracted parenting. In later life, facing an enforced retirement from parenthood, they may refuse the "feminine" alternatives represented by Passive Mastery, and instead continue to maintain their denials in the more extreme form of Magical Mastery; that is, they may elect to ignore the troubling realities of aging that they can no longer avoid or change through direct action.

Chapter 9

1. While I have noted many instances of matriarchal governance in extradomestic settings, the old woman's rule is confined mainly to the home. Thus, in his cross-cultural review of aging, Simmons finds women, whether young or old, holding high office in only two out of seventy-one preliterate societies. He does, however, note that older women become increasingly important in managing the initiatory rites of young people. His general conclusion is that while older men may be inherently valued in public life, older women are not: "Wherever women have been respected in old age, men are rarely without honor; but prestige for aged men has been no assurance of similar status for aged women."

2. A study by Barker and Barker (1961) underlines the tendency of the aged to move toward the "ritual track," even in relatively nontraditional, "modern" settings, so long

Notes

as a vestige of the religious option is made available to them. The authors identified the various "behavior settings" of two small communities, one in the midwestern United States and the other in England, and found that "the old people of both communities overinhabit behavior settings where religion and art are prominent and underinhabit settings where education, physical health, government, social interaction, and recreation are prominent." In effect, the aged of both societies have shifted their investment from those behavior settings that stress physical activity, competitive political engagement, and authority over the young, in favor of settings that stress submission to a higher sacred authority.

3. While LeVine (1976) asserts that older men move to the ritual track in order to compensate for failures in their productive and political life, Kracke (1977) disputes this: "The spiritual realm is not, as it is sometimes pictured, an avenue to self-esteem for those unable to obtain political power. It is a separate realm of at least equal significance in its own right."

4. This pattern repeats: Halfway around the world, the future medicine man of the Sioux moved, via the vision quest, through privation to power. The Sioux candidate went into the desert without food, clothing, or arms. He carried with him nothing of society's protection. He starved, he froze at night, he sang his death song. Finally, when he was delirious, his totem animal appeared to him in a vision and indicated to him the source of his future power. Returning to the tribe, the candidate knew his totem lodge and the appropriate rituals that would link him with the power of his totemic sponsor.

Consider also Róheim's description of the magician's initiation in the Australian bush tribes. He goes to the place of the spirits; he lies down passively, facing toward their cave; he is taken into their cave, cut to pieces, eaten, and put together again with a new set of organs by the same spirits that destroyed him. He emerges from the cave imbued with the power of the destroying spirits who now have become his sponsors. In effect, he has gone through a cycle of death and rebirth. Having invited the ultimate passivity, and having survived the fearful contact with the spirits, he has beaten death and has thereby transcended the final and greatest of human limitations. He has some of the omnipotence of the immortal gods.

5. Other examples, drawn from around the world, attest to the near divine powers of the aged, and particularly older men. In the Pacific region, the Kowranega of Australia call their older men by a name akin to "superman." And on the northern edge of the Pacific basin, the Bering Eskimo assign the elders of their "patriclans" a crucial leadership role: They alone have the mighty aura of age and can be the keepers of the religious lore. Among these Eskimo societies, the elders, by virtue of their contact with the spirit world, have the unique task of naming a newborn child. The neonate is regarded as the temporary tenement of an Eskimo immortal represented by the "name-soul," and it is only the aged who are strong enough and close enough to the spirit world to dare the contact with the immortals and to inquire if they are ready for reincarnation.

Among the Eskimo, as with other groups, the strength of the aged is special, and does not betoken the pragmatic, physical sources from which young men mainly draw their strength. Thus, in many Eskimo societies, young men could not acquire magic spells from the aged until their own physical strength had begun to wane.

Spelling out this special linkage between seniority and the supernatural, Simmons reports that among the Arunta of Australia age and magic combined were far more effective than sheer physical prowess. Moreover, senescent mana is not confined to the older man's body parts but dominates a zone of dangerous influence around him; Cowgill (1968) tells us that younger Thai will crouch before their elders in order to avoid intruding into the zone of *tabu* power that extends outward from the head of the senior person.

The special mana of the elderly is so great that it can even harm its possessor. According to Janice Hurlbert (1962), the Hare Indians of the northwest territories of Canada believe that a medicine man's power will increase to the point that it blinds him as he ages. And if the special powers of the old man are sometimes a threat to him, they also, in the primitive setting, make him invulnerable to *tabu* influences that could threaten the young. We have already cited many cultural instances in which the choicest foods were made taboo for the young and were reserved for the old. While Simmons implies that the aged have bamboozled the young into believing that the best

313

foods would be bad for them, this commonsense interpretation does not hold up. Earlier, as we reviewed the reasons behind the age-graded food taboos of various cultures, we found a common theme: Food contains *tabu* power; as such, it is double-edged, holding the virtue either to build or to destroy. Only the older person, who has survived the debilities that come with aging, or who has counterpower—mana—of his own, can risk the destructive potential inherent in such mana-charged foods.

6. We should not be surprised to find that, in the traditional society, the emergent propensities of later life mortise into those age-specific roles that both require them and sponsor their further development, as identity resources. By far the longest period of human species existence was passed in isolate, preliterate, traditionally oriented communities. These were the species-specific settings in which human characteristics evolved, particularly those ego executive capacities that organize personal-communal relationships. The personal-social integrations that we find among older traditionals are not accidental, and they were, in human prehistory, the rule rather than the exception. They may register a process of interlocking evolution that over the millennia selected for viable human traits and for viable institutions—for roles and conventions that lifted crude human potentials to the level of communication and social utility. Thus the study of the old traditionals may constitute a kind of social archaeology. They may reveal to us the close, almost umbilical connection between man and the folk community that was once the general human condition.

7. In addition, for both mothers and children, the extended family provides a special ecology for dealing with the ambivalent feelings that accompany all intimate relationships and that can poison the climate of the indrawn, isolated nuclear family. In the nuclear family, we often hate those that we depend on—those, paradoxically, that we also love. Intrapersonal and interpersonal conflict is unavoidable, destabilizing nuclear family and personality systems alike; but in the extended family, which routinely provides a panel of kinsmen who hold the title and role of "mother," "father," "brother," and "sister," the elements of love and hate can be divided from each other and allocated toward separate parental and sibling figures. The extended family can generate (and amplify) feuds between clans, but it tends to mitigate inner, neurotic conflicts within growing children.

Chapter 10

1. Modernization and urbanization, though they can be part of a common sequence, are often independent of each other. A society can move toward modernity without acquiring an urban aspect, and cities (Mecca, for example) can be the headquarters of a traditional way of life without experiencing modernity. Briefly, modernity refers to influences that, while they usually originate in cities, are not exclusive to them: literacy, electrification, advanced modes of technology and manufacture, as well as "rationalized" uses of labor. Urbanization refers to the forms of association that typically occur in the city but rarely in the village: the mingling of foreigners; the mixing of social classes and their contrasting ways; the display of various philosophies—the city as marketplace of lifeways, ideas, and goods.

2. This shift in the bases of financial power has similar effects in Africa. In one of the few cross-cultural studies linking mental illness to loss of elderly prestige, Malcolm Arth (1968) finds that young Ibo, who now acquire independent wealth from wage work, no longer rely on their fathers for the bride price. In consequence, the father's prestige declines, and the author finds that "psychosenility" is beginning to appear among the elderly casualties of Ibo modernization.

3. Levy (1967) found that the government's stock-reduction policy (instituted to counter overgrazing) motivates older Navajo to surrender their grazing permits—and the associated prestige—to the young in exchange for welfare benefits. The Navajo religion is based on healing rituals known mainly to older traditionals, but now the singer's role is declining in competition with government hospitals and with the Peyote religion, which makes ceremonial roles available at low cost to young Navajo. Furthermore, the traditionalist loses his educational role; his grandchildren now go to

Notes

Indian Agency boarding schools and learn "white man" knowledge that he lacks. Finally, the old Navajo becomes a source of strife among clan members who compete for his custody in order to get his welfare check. The old-age assistance program was designed to safeguard the welfare and prestige of the aging Indian; instead, it gives him the image of a dependent rather than a wise man.

4. The psychosomatic outcomes of modernization are among the most striking. A Druze doctor, practicing in Majd-Al-Shams, told me that the oldest men, perhaps buffered by their religious affiliations, are not likely to suffer from stress-related diseases; rheumatism and the various effects of being overweight are the chief complaints of the old *aqil*. There are only a few cases of cardiovascular disorder among the older generation who do tend not to suffer from drawn-out degenerative disease, but rather die suddenly, usually in their late seventies, from a stroke or a massive heart attack. In fact, chronic heart disease is more prevalent among the secularized middle-aged—those who drink and smoke, and who are not folded into the traditional religion—than among the religious aged. Other stress diseases, such as ulcers, similar gastrointestinal disorders, and hypertension, are also more common among the younger adults, who are more open to and vulnerable to the enticements and pressures of the modern, extra-Druze world.

The same effect can be observed along the rural/traditional-urban/secular continuum of the Navajo: For any age group, the incidence of stress-related Navajo disease is always lower among the traditional Highlanders. The incidence of mental illness, of stress-related somatic disease, and of the alcoholism that promotes such disease, rises sharply as one moves downslope to Tuba City, the Western administrative center for the Navajo.

5. This implication, that the dehumanizing pressures of the city impact more heavily on older men than on older women, is borne out by the mortality statistics. Gerrit Kooy (1962) discovers that, except in rural areas, older Dutch women outnumber older Dutch men. By the same token, Chevry (1962) reports that the proportion of aged men, particularly widowers, increases in step with the agricultural character of the French municipality. And comparisons by Jacqueline Jackson (1971) of rural and urban American blacks also suggest a correlation between later-life survival and the nature of familial bonds. Rural black men outlive their women, but the urban shift toward superior female longevity is accompanied by a corresponding shift in the patterning of affectional ties. In the city, these are strongest between mothers and daughters, and this close female bond pays off in the form of greater direct assistance to the aging black mother at all social class levels.

In this vein, Ethel Shanas et al. (1968) find that modified extended family networks are common in all urban centers of the industrialized West. For the elderly widow, disengagement is not obligatory; bereavement draws reintegrating responses from separately domiciled family members who still maintain strong mutual ties. Citing his Austrian studies, Leopold Rosenmayer (1973) terms this pattern "intimacy at a distance." Townsend (1962) notes the "repair" function of this family format: Since urban women typically outlive men, the modified extended family helps to compensate the English widow for the loss of her spouse.

6. Urban male and female parents seem more ready than rural parents to conserve for themselves rather than conceding to children the omnipotentiality that is the keynote of the preparental (and postparental) stages of life. The cities sponsor the psychology of self-seeking individualism, which advances commerce and the lively arts; but the egocentric psychology that is vital to creativity is also at odds with the requirements of reliable parenting. Small wonder, then, that across history the rural sector of society has shown much more staying power than the urban sector. Cities rise and fall, moving quickly from hectic growth to decadence and decay; but the rural sector, which preserves proparental attitudes, underwrites thereby its own continuity.

7. Colin Turnbull's (1972) powerful portrayal of the Ik tribe of Kenya provides a chilling example of the human disaster that ensues when society loses culture. The Ik had recently lost their ancient hunting preserves, and this dispossession invalidated the mythic sources of their culture; in effect, if the gods could not protect them, then the gods were dead, and with them died the cultural principles that the gods had sacralized and legitimized. When the cultural rules were no longer idealized, they were

at best only weakly internalized, and they lost their power to bind narcissism and aggression to the social weal. In consequence, unmodified narcissism is the major theme of human relationships among the Ik, to the point where children are literally evicted from the parental home at the age of three. If they are lucky, they may become part of a juvenile food-gathering band and survive to adulthood. Predictably, the fate of the relatively useless Ik aged parallels that of the children. They have no honor, no secure place within ritual or within the family, and they are left to die when they can no longer fend for themselves.

BIBLIOGRAPHY

Abarbanel, J. (1971). *Aging, family phase, and intergenerational relations in an Israeli farming cooperative.* Paper presented at the December 1972 meetings of the American Gerontological Society, San Juan, PR.

Abraham, K. (1966). *On character and libido development: Six essays* (B. Lewin, Ed.). New York: W. W. Norton.

Achenbaum, W. A. (1974). *Old age in America.* Unpublished doctoral dissertation, University of Michigan, Ann Arbor.

Amoss, P. T. (1981). Coast Salish elders. In P. T. Amoss and S. Harrell (Eds.), *Other ways of growing old: An anthropological perspective* (pp. 227–247). Stanford: Stanford University Press.

Arensberg, C. M., and Kimball, S. T. (1968). *Family and community in Ireland.* Cambridge: Harvard University Press.

Arth, M. (1968). Ideals and behavior: A comment on Ibo respect patterns. *Gerontologist, 8,* 242–244.

Atchley, R. (1976). Selected social and psychological differences between men and women in later life. *Journal of Geronotology, 31*(2), 204–211.

Barker, R. G., and Barker, L. S. (1961). The psychological ecology of old people in Midwest, Kansas, and Yoredale, Yorkshire. In B. Neugarten (1975), *Middle age and aging* (pp. 453–460). Chicago: University of Chicago Press.

Barry, H., Bacon, M., and Child, I. (1957). A cross-cultural survey of some sex differences in socialization. *Journal of Social and Abnormal Psychology, 55,* 327–432.

Batchelor, J. (1928). *The Ainu of Japan.* New York: Fleming Revell Company.

Benedek, T. (1952). *Psychosexual functions in women.* New York: Ronald Press.

Berezin, M. (1963). Some intra-psychic aspects of aging. In N. Zinberg (Ed.), *Normal psychology of the aging process* (pp. 17–71). New York: International Universities Press.

Bergler, R. (1961). Beitrage zur Psychologie des Erwachsen en Alters. In *Bibliotheca vita humana.* Basel: Karger.

Best, E. (1925). *The Maori* (Vol. 1). Wellington, New Zealand: Board of Maori Ethnological Research.

317

Bibliography

Bischofsberger, O. (1972). *The generation classes of the Zanaki (Tanzania)*. Fribourg: University Press.

Brenneis, C. B. (1975). Developmental aspects of aging in women. *Archives of General Psychiatry, 32*, 429–435.

Brown, J. (1985). *In her prime: A new view of middle-aged women*. South Hadley, MA: Bergin and Garvey Publishers, Inc.

Bunzel, R. (1930). *Introduction to Zuñi ceremonialism* (47th Annual Report). Washington, DC: Bureau of American Ethnology.

Burgess, E. (1960). Aging in western culture, In E. Burgess (Ed.), *Aging in western societies* (pp. 28–30). Chicago: University of Chicago Press.

Cameron P. (1967). Introversion and egocentricity among the aged. *Journal of Gerontology, 22*, 465–468.

Cesa-Bianchi, M. (1962). A further contribution to the study of adjustment in old age. In C. Tibbitts and W. Donahue (Eds.), *Social and psychological aspects of aging: Aging around the world* (pp. 523–627). New York: Columbia University Press.

Chagnon, N. A. (1968). *Yanomamo: The fierce people*. New York: Holt, Rinehart and Winston.

Chevry, G. (1962). One aspect of the problem of older persons: Housing conditions. In C. Tibbitts and W. Donahue (Eds.), *Social and psychological aspects of aging: Aging around the world* (pp. 98–110). New York: Columbia University Press.

Clancy, K. L. (1975). Preliminary observations on media use and food habits of the elderly. *Gerontologist, 15*, 529–532.

Colson, E., and Scudder, T. (1981). Old age in Gwembe District, Zambia. In P. T. Amoss and S. Harrell (Eds.), *Other ways of growing old: Anthropological perspectives* (pp. 125–153). Stanford: Stanford University Press.

Cooper, K. L., and Gutmann, D. L. (1987). Gender identity and ego mastery style in middle-aged, pre– and post–empty nest women. *Gerontologist, 27*(3), 347–352.

Cowgill, D. (1968). The social life of the aging in Thailand. *Gerontologist, 8*, 159–163.

Cox, F. M. (1977). *Aging in a changing village society: A Kenyan experience*. Washington, DC: International Federation of Aging.

Cronin, J. (1982). *Attorneys' perceptions of their roles in divorce proceedings*. Unpublished doctoral dissertation, Northwestern University Medical School, Department of Psychiatry, Chicago.

Cumming, E., and Henry, W. (1961). *Growing old: The process of disengagement*. New York: Basic Books.

de Beauvoir, S. (1972). *The coming of age*. New York: J. Putnam's Sons.

Elwin, V. (1958). *Myths of the north-east frontier of India*. Shillong, India: North-East Frontier Agency.

Ewing, K. P. (1981). *The crisis of marriage for two Pakistani women*. Unpublished manuscript, University of Chicago, Department of Anthropology.

Feldman, S., Biringen, C., and Nash, S. (1981). Fluctuations of sex-related self-attributions as a function of stage of family life cycle. *Developmental Psychology, 17*, 24–35.

Fenichel, O. (1942). *The psychoanalytic theory of neurosis*. New York: W. W. Norton.

Ferenczi, S. (1913). *Further contributions to the theory and technique of psychoanalysis*. London: Hogarth Press.

Fortes, M. (1949). *The web of kinship among the Tallensi*. London: Oxford University Press.

Freud, S. (1911). Psychoanalytic notes upon an autobiographical account of a case of paranoia (dementia paranoides). Republished in 1958, in J. Strachey (Ed. and Trans.), *The standard edition of the complete psychological works of Sigmund Freud* (Vol. 12, pp. 9–82). London: Hogarth Press.

Freud, S. (1913). Totem and taboo. Republished in 1938, in A. A. Brill (Ed.), *The basic writings of Sigmund Freud*. New York: Modern Library.

Galler, S. (1977). *Women graduate student returnees and their husbands: A study of the effects of the professional and academic graduate school experience on sex-role perceptions, marital relationships, and family concepts.* Unpublished doctoral dissertation, Northwestern University, School of Education, Evanston, IL.

Gelles, R. (1979). *Family violence.* Beverly Hills: Sage Publications.

Giambra, L. (1973). Daydreaming in males from seventeen to seventy-seven: A preliminary report. *Proceedings of the 81st Annual Convention of the American Psychological Association, 8,* 769–770.

Gilder, G. (1973). *Sexual suicide.* New York: Quadrangle Press.

Gluckman, M. (1955). *Custom and conflict in Africa.* Glencoe, IL: Free Press.

Gold, S. (1960). Cross-cultural comparisons of role change with aging. *Student Journal of Human Development* [University of Chicago], *1,* 11–15.

Goody, J. (1962). *Death, property and the ancestors.* Stanford: Stanford University Press.

Grattan, C. (1948). *An introduction to Samoan custom.* Apia, Western Samoa: Samoa Printing and Publishing Co.

Griffin, B. (1984). *Age differences in preferences for continuity and change.* Unpublished doctoral dissertation, Northwestern University Medical School, Division of Psychology, Department of Psychiatry and Behavioral Sciences, Chicago.

Gutmann, D. L. (1964). An exploration of ego configurations in middle and later life. In B. Neugarten (Ed.), *Personality and later life* (pp. 114–148). New York: Atherton.

Gutmann, D. L. (1968). Aging among the Highland Maya: A comparative study. *Journal of Personality and Social Psychology, 7,* 28–35.

Gutmann, D. L. (1969). The country of old men: Cross-cultural studies in the psychology of later life. In *Occasional papers in gerontology.* Ann Arbor: University of Michigan Press.

Gutmann, D. L. (1974). Alternatives to disengagement: Aging among the Highland Druze. In R. LeVine (Ed.), *Culture and personality: Contemporary readings* (pp. 232–245). Chicago: Aldine.

Guttman, D. (1973). Leisure-time activity interests of Jewish aged. *Gerontologist, 13*(2), 219–223.

Harlan, W. (1964). Social status of the aged in three Indian villages. *Vita Humana, 7,* 239–252.

Harlow, H., and Mears, C. (1979). *The human model: Primate perspectives.* Washington, DC: V. H. Winston.

Harrell, S. (1981). Growing old in rural Taiwan. In P. T. Amoss and S. Harrell (Eds.), *Other ways of growing old: Anthropological perspectives* (pp. 193–210). Stanford: Stanford University Press.

Havighurst, R. (1960). Life beyond family and work. In E. Burgess (Ed.), *Aging in Western societies* (pp. 299–353). Chicago: University of Chicago Press.

Hayes, R. O. (1975). Female genital mutilation, fertility control, women's roles, and the patrilineage in modern Sudan: A functional analysis. *American Ethnologist, 2,* 617–633.

Heath, D. H. (1972). What meaning and effect does fatherhood have for the maturing professional man? *Merrill-Palmer Quarterly, 24,* 125–133.

Heron, A., and Chown, S. (1962). Semi-skilled and over forty. In C. Tibbitts and W. Donahue (Eds.), *Social and psychological aspects of aging: Aging around the world* (pp. 195–207). New York: Columbia University Press.

Hiebert, P. G. (1981). Old age in a south Indian village. In P. T. Amoss and S. Harrell (Eds.), *Other ways of growing old: Anthropological perspectives* (pp. 211–226). Stanford: Stanford University Press.

Hoffman, M. L. (1970). Conscience, personality and socialization techniques. *Human Development, 13,* 90–126.

Holmberg, A. (1961). Age in the Andes. In R. Kleemeier (Ed.), *Aging and Leisure* (pp. 86–90). New York: Oxford University Press.

Honigmann, J. J. (1954). *Culture and personality.* New York: Harper and Row.

319

Howell, S. C., and Loeb, M. B. (1969). Nutrition and aging: A monograph for practitioners. *Gerontologist*, 9(3: Pt. 3), 7–122.

Hrdy, S. B. (1981). "Nepotists" and "altruists": The behavior of old females among macaques and langur monkeys. In P. T. Amoss and S. Harrell (Eds.), *Other ways of growing old: Anthropological perspectives* (pp. 59–76). Stanford: Stanford University Press.

Hughes, C. (1960). *An Eskimo village in the modern world*. Ithaca, NY: Cornell University Press.

Hurlbert, J. (1962). *Age as a factor in the social organization of the Hare Indians of Fort Good Hope, Northwest Territories*. Ottawa, Ontario, Canada: Northern Coordination and Research Centre, Department of Northern Affairs and National Resources.

Ikels, C. (1975). Old age in Hong Kong. *Gerontologist*, 15, 230–235.

Jackson, J. (1971). Sex and social class variations in black aged parent-adult child relationships. *Aging and Human Development*, 2, 96–107.

Jacobowitz, J. (1984). *Stability and change of coping patterns during the middle years as a function of personality type*. Unpublished doctoral dissertation, Hebrew University at Jerusalem, Department of Psychology.

Jaslow, P. (1976). Employment, retirement and morale among older women. *Journal of Gerontology*, 31(2), 212–218.

Jung, C. G. (1933). *Modern man in search of a soul*. New York: Harcourt, Brace and World.

Kardiner, A., and Linton, R. (1945). *The psychological frontiers of society*. New York: Columbia University Press.

Kaufman, M. (1940). Old age and aging: The psychoanalytic point of view. *American Journal of Orthopsychiatry*, 10, 73–89.

Keith-Ross, J. (1974). Life goes on: Social organization in a French retirement residence. In J. F. Gubrium (Ed.), *Late life: Recent readings in the sociology of aging*. Springfield, IL: Charles C. Thomas.

Kelly, E. L. (1955). Consistency of the adult personality. *American Psychologist*, 10, 659–681.

Kerchoff, A. (1966). Family patterns and morale. In I. Simpson and J. McKinney (Eds.), *Social aspects of aging* (pp. 173–192). Durham, N.C.: Duke University Press.

Kleemeier, R. (1960). The mental health of aging. In E. Burgess (Ed.), *Aging in Western societies* (pp. 203–270). Chicago: University of Chicago Press.

Kooy, G. (1962). The aged in the rural Netherlands. In C. Tibbitts and W. Donahue (Eds.), *Social and psychological aspects of aging: Aging around the world* (pp. 501–509). New York: Columbia University Press.

Kracke, W. H. (1977). *Some frequent life crises in Kagwahiv adulthood*. Paper given at the symposium on the Cultural Phenomenology of Adulthood and Aging, Cambridge, MA.

Krohn, A., and Gutmann, D. L. (1971). Changes in mastery style with age: A study of Navajo dreams. *Psychiatry*, 34, 289–300.

Kubey, R. (1980). Television and aging: Past, present and future. *Gerontologist*, 20, 16–35.

Kupper, G. (1975). *Exploring the impact of parenthood on young college students*. Unpublished honors thesis, University of Michigan, Ann Arbor.

Lambeck, M. (1985). Motherhood and other careers in Mayotte (Comoro Islands). In J. K. Brown and V. Kerns (Eds.), *In her prime: A new view of middle-aged women* (pp. 67–84). South Hadley, MA: Bergin and Garvey.

Lantis, M. (1953). The Nunivak Eskimo of the Bering Sea. In I. T. Sanders, *Societies around the world* (pp. 74–99). New York: Dryden Press.

Leonard, O. (1967). The older Spanish-speaking people of the Southwest. In E. Youmans

Bibliography

(Ed.), *The older rural Americans* (pp. 239–261). Lexington: University of Kentucky Press.

LeVine, R. (1963). Witchcraft and sorcery in a Gusii community. In J. Middleton and E. Winter (Eds.), *Witchcraft and sorcery in East Africa* (pp. 113–128). London: Routledge and Kegan Paul.

LeVine, R. (1976). *Intergenerational tensions and extended family structures in Africa.* Unpublished manuscript, University of Chicago, Committee on Human Development.

Levy, J. (1967). The older American Indian. In E. Youmans (Ed.), *The older rural Americans* (pp. 231–238). Lexington: University of Kentucky Press.

Levy, R. (1977). *Notes on being adult in different places.* Unpublished paper, University of California, San Diego, Department of Anthropology.

Lévy-Bruhl, L. (1928). *The "soul" of the primitive.* London: Allen and Unwin.

Lewis, O. (1951). *Life in a Mexican village: Tepoztlan restudied.* Urbana: University of Illinois Press.

Lindholm, C. (1981). The structure of violence among the Swat Pakhtun. *Ethnology*, 20(2), 147–156.

Lovald, K. (1962). The social life of the aged on skid row. In C. Tibbitts and W. Donahue (Eds.), *Social and psychological aspects of aging: Aging around the world* (pp. 510–517). New York: Columbia University Press.

Lowenthal, M., Thurnher, M., and Chiriboga, D. (1975). *Four stages of life.* San Francisco: Jossey-Bass.

McCrae, R., and Costa, P. T. (1984). *Emerging lives, enduring dispositions: Personality in adulthood.* Boston: Little, Brown.

Mandelbaum, D. (1957). The world and the world view of the Kota. In M. Marriott (Ed.), *Village India* (pp. 219–236). Chicago: University of Chicago Press.

Mead, M. (1928). *Coming of age in Samoa.* New York: William Morrow.

Mead, M. (1966). *New lives for old: Cultural transformation—Manus, 1928–1953.* New York: William Morrow.

Mead, M. (1967). Ethnological aspects of aging. *Psychosomatics*, 8(supp.), 33–37.

Meerloo, J. (1955). Psychotherapy with older people. *Geriatrics*, 10, 583–590.

Menaghan, E. G. (1975). *Parenthood and life satisfaction in later life: A comparative analysis.* Unpublished paper, University of Chicago, Committee on Human Development.

Mernissi, F. (1975). *Beyond the veil: Male-female dynamics in modern Muslim society.* Cambridge, MA: Schenkman Publishing Company.

Middleton, J. (1953). *The central tribes of the north-eastern Bantu.* London: International African Institute.

Morsy, S. A. (1978). *Gender, power and illness in an Egyptian village.* Unpublished doctoral dissertation, Michigan State University, Lansing.

Murdoch, G. (1935). Comparative data on the division of labor by sex. *Social Forces*, 15, 551–553.

Nadel, S. F. (1952). Witchcraft in four African societies: An essay in comparison. *American Anthropologist*, 54, 18–29.

Neugarten, B., Havighurst, R., and Tobin, S. (1968). Disengagement and patterns of aging. In B. Neugarten (Ed.), *Middle age and aging: A reader in social psychology* (pp. 161–172). Chicago: University of Chicago Press.

Newton, N. (1973). *Psycho-social aspects of the mother/father/child unit.* Paper presented at meetings of the Swedish Nutrition Foundation, Upsala.

O'Connel, L. (1981). *Late parenthood and personality change in middle-aged men: An examination of the developmental hypothesis for adulthood.* Unpublished doctoral dissertation, University of Chicago, Committee on Human Development.

Okada, Y. (1962). The aged in rural and urban Japan. In C. Tibbitts and W. Donahue

(Eds.), *Social and psychological aspects of aging: Aging around the world* (pp. 454–458). New York: Columbia University Press.

Partch, J. (1978). *The socializing role of postreproductive Rhesus macaque females.* Paper presented at meetings of the American Association of Physical Anthropologists, Toronto, Ontario, Canada.

Paulig, K., and McGee, J. *Situational variation and sex differentiation in group interaction patterns of the elderly.* Paper presented at the November 1977 meetings of the Gerontological Society, San Francisco.

Perloff, R., and Lamb, M. (1980). *The development of gender roles: An integrative lifespan perspective.* Unpublished manuscript, University of Wisconsin, Madison, Department of Psychology.

Peskin, H., and Livson, N. (1981). Uses of the past in adult psychological health. In D. Eichorn, P. Mussen, D. Clausen, R. Hann, and C. Honzik (Eds.), *Present and past in middle life* (pp. 153–181). New York: Academic Press.

Peterson, R. (1979). The relationship of middle-aged children and their parents. In P. K. Ragan (Ed.), *Aging parents* (pp. 22–37). Los Angeles: University of Southern California Press.

Pospisal, L. (1958). *Kapauka Papuans and their law.* New Haven: Yale University Publications in Anthropology, No. 54.

Powers, S. (1877). Tribes of California. In *Contributions to North American ethnology*, Vol. 3. Washington, DC: Department of the Interior.

Press, I., and McKool, M. (1972). Social structure and status of the aged: Toward some valid cross-cultural generalizations. *Aging and Human Development, 3*, 297–306.

Prins, A. H. J. (1953). *East African age-class systems.* Groningen, W. Ger.: J. B. Wolters.

Purifoy, F. W., Koopmans, L. H., and Tatum, R. W. (1980). Steroid hormones and aging: Free testosterone, testosterone and androstenedione in normal females aged 20–87 years. *Human Biology, 52*(2), 181–191.

Quain, B. (1948). *Fijian village.* Chicago: University of Chicago Press.

Redfield, R. (1947). The folk society. *American Journal of Sociology, 2*(4), 293–308.

Richards, A. I. 1956. *Chisungu.* New York: Grove Press.

Ripley, D. (1984). *Parental status, sex roles, and gender mastery style in working class fathers.* Unpublished doctoral dissertation, Illinois Institute of Technology, Department of Psychology, Chicago, Illinois.

Róheim, G. (1930). *Animism, magic and the divine king.* New York: Knopf.

Rose, A. M. (1960). The impact of aging on voluntary associations. In C. Tibbitts (Ed.), *Handbook of social gerontology.* Chicago: University of Chicago Press.

Rosenmayr, L. (1973). Family relations of the elderly. *Zeitschrift fur Gerontologie, 6,* 272–283.

Rowe, W. (1961). The middle and later years in Indian society. In R. Kleemeier (Ed.), *Aging and leisure* (pp. 104–112). New York: Oxford University Press.

Roy, M. (1975). *Bengali women.* Chicago: University of Chicago Press.

Rudolph, S., and Rudolph, L. (1978). Rajput adulthood: Reflections on the Amar Singh diary. In E. H. Erickson (Ed.), *Adulthood* (pp. 149–170). New York: W. W. Norton.

Rustom, C. (1961). The aging Burman. In R. Kleemeier (Ed.), *Aging and leisure* (pp. 100–103). New York: Oxford University Press.

Shanan, J. (1985). Personality types and culture in late adulthood. In J. Meacham (Ed.), *Contributions to human development, 12,* 1–144. Basel: S. Karger.

Shanas, E., Townsend, P., Wedderburn, D., Friis, H., Milhøj, P., Stehouwer, J. (1968). *Old people in three industrial societies.* New York: Atherton Press.

Sharp, H. S. (1981). Old age among the Chipewyan. In P. T. Amoss and S. Harrell (Eds.), *Other ways of growing old: Anthropological perspectives* (pp. 99–109). Stanford: Stanford University Press.

Sheehan, T. (1976). Senior esteem as a factor of socio-economic complexity. *Gerontologist, 16,* 433–440.

Bibliography

Shelton, A. (1965). Ibo aging and eldership: Notes for gerontologists and others. *Gerontologist, 5,* 20–23.

Simmons, L. W. (1945). *The role of the aged in primitive society.* New Haven: Yale University Press.

Smith, R. (1961). Japan: The later years of life and the concept of time. In R. Kleemeier (Ed.), *Aging and leisure* (pp. 84–112). New York: Oxford University Press.

Spencer, P. (1965). *The Samburu: A study of gerontocracy in a nomadic tribe.* Berkeley: University of California Press.

Spitz, R. A., and Wolf, K. M. (1946). Anaclitic depression. *Psychoanalytic Study of the Child, 22,* 313–342. New York: International Universities Press.

Steed, G. (1957). Notes on an approach to personality formation in a Hindu village in Gujarat. In M. Marriott (Ed.), *Village India* (pp. 102–144). Chicago: University of Chicago Press.

Streib, G. (1968). Family patterns in retirement. In M. S. Sussman (Ed.), *Sourcebook in marriage and the family.* Boston: Houghton Mifflin.

Strong, E. L. (1931). *Changes of interests with age.* Stanford: Stanford University Press.

Swartley, A. (1986). Life after motherhood. *Village Voice Literary Supplement,* June issue, 9–11.

Tachibana, K. (1962). A study of introversion-extraversion in the aged. In C. Tibbitts and W. Donahue (Eds.), *Social and psychological aspects of aging: Aging around the world* (pp. 655–656). New York: Columbia University Press.

Talmon-Garber, Y. (1962). Aging in collective settlements in Israel. In C. Tibbits and W. Donahue (Eds.), *Social and psychological aspects of aging: Aging around the world* (pp. 426–441). New York: Columbia University Press.

Thomae, H. (1962). Thematic analysis of aging. In C. Tibbitts and W. Donahue (Eds.), *Social and psychological aspects of aging: Aging around the world* (pp. 657–663). New York: Columbia University Press.

Townsend, P. (1957). *The family life of old people.* Glencoe, IL: Free Press.

Townsend, P. (1962). The purpose of the institution. In C. Tibbitts and W. Donahue (Eds.), *Social and psychological aspects of aging: Aging around the world* (pp. 378–399). New York: Columbia University Press.

Turnbull, C. (1972). *The mountain people.* New York: Simon and Schuster.

Van Arsdale, P. W. (1981). The elderly Asmat of New Guinea. In P. T. Amoss and S. Harrell (Eds.), *Other ways of growing old: Anthropological perspectives* (pp. 111–123). Stanford: Stanford University Press.

Vatuk, S. (1975). The aging woman in India: Self-perceptions and changing roles. In A. DeSouza (Ed.), *Women in contemporary India* (pp. 143–163). Delhi: Manohar.

Webber, I., Coombs, D., and Hollingsworth, J. (1974). Variations in value orientations by age in a developing society. *Journal of Gerontology, 29,* 676–683.

Wershow, H. J. (1969). Aging in the Israeli kibbutz. *Gerontologist, 9,* 300–304.

Westermarck, E. (1926). *Ritual and belief in Morocco.* 2 vols. London: Kegan Paul.

Williams, T., and Calvert, J. (1859). *Fiji and the Fijians.* New York: Appleton and Co.

Wolf, M. (1974). Chinese women: Old skills in a new context. In M. Z. Rosaldo and L. Lamphere (Eds.), *Woman, culture and society* (pp. 157–172). Stanford: Stanford University Press.

Wolff, K. (1959). *The biological, sociological and psychological aspects of aging.* Springfield, IL: Charles C. Thomas.

Yang, C. K. (1959). *Chinese communist society: The family and the village.* Cambridge, MA: MIT Press.

Yap, P. (1962). Aging in under-developed Asian countries. In C. Tibbitts and W. Donahue (Eds.), *Social and psychological aspects of aging: Aging around the world* (pp. 442–453). New York: Columbia University Press.

Youmans, E. (1967). Disengagement among older rural and urban men. In E. Youmans

(Ed.), *The older rural Americans* (pp. 97–116). Lexington: University of Kentucky Press.

Young, M., and Geertz, H. (1961). Old age in London and San Francisco: Some families compared. *British Journal of Sociology, 12*(2), 124–141.

Zinberg, N., and Kaufman, I. (1963). Cultural and personality factors associated with aging: An introduction. In N. Zinberg and I. Kaufman (Eds.), *Normal psychology of the aging process* (pp. 17–71). New York: International Universities Press.

INDEX

Abarbanel, J., 93

Abraham, K., 130

Abstinence, from sex, 91

Accommodation, in older men, 25–26; Passive Mastery and, 51, 65

Achenbaum, W. A., 240

Action, gender of, 79–81

Active coping, 68

Adolescence, 3

Adoption, 179

Adultery, 162; tribal society and, 85

Adulthood: major events during, 44; parenthood and, 187, procreative period of, 6

Afghanistan, tribal society in, 166

Africa, tribal society in, 85–86, 89, 162, 169, 176, 314–16

Age-grading systems, 84–92; female dominance and, 155–60, 166–84; universal themes in, 83

Aggression: aging men and, 94–97; bargaining and, 48; female, 24, 47; flamboyant, 94; male, age and social regulation of, 82–97; male, general issues in, 292–95

Aging: agriculture and, 3–4; cultural phobia against, 4–5, 8–9, 18–19; decreased intellectual sharpness as consequence of, 8; as depletion, 4–5; egocentricity and, 89, 97, ethnography and study of, 21; feminization and, 88; gerontological emphasis on masculine aspects of, 38; inferiority feelings and, 97; mana and, 220–22, 227, 313–14; medicine and, 3–4; necessity for in evolutionary sense, 4, 185; orality and, 103–18, 121, 125–26, 129–31; parental imperative and, 22, 185–213; routes to pleasure and, 99–103; self-sufficiency and, 161–62; tangible changes of, 8; *see also* Elderhood; Elders; Men, older; Women, older

Alcohol abuse: by Highland Maya, 39, 118, 258–61; lack of, by Druze, 111; by Navajo, 110, 254, 258–59; Passive Mastery and, 61

Allah, Druze and, 36–37, 42, 92, 222–28, 250, 286–87, 291

Altruism: maternal, 303–4; parental, 232

Amoss, P. T., 160

Androgens, 181

Androgyny, 74, 135, 184, 212, 216–17, 224, 233

Animals, lower, 6

325

Index

Index

Life-cycle: differences, 57; social reg-
ulation of males across, 155; staging
of, 83
Life Satisfaction Scale, 145–47, 304
Lindholm, C., 166
Linton, R., 10–13, 15–20, 41, 44, 88, 160
Linton levels: first, 10–11, 16–17, 20, 41,
44, 231, 236; second, 11–12, 15–16, 19, 41,
226–27, 231, 236; third, 12–13, 16, 18–19,
41, 44, 226, 231, 236
Livson, N., 210
Loeb, M., 117
Longevity, 3–4, 6; cultural differences
in, 116; geographic isolation and, 4;
orality and, 126; of women over
men, 244
Lovald, K., 307
Love, 102
Lowenthal, M., 156–57

McCrae, R., 70
McGee, J., 158
Machismo: social terminus of, 88; of
younger men, 95
McKool, M., 237
Mandelbaum, D., 188
Manson, Charles, 253
Marriage, 171–72, 188, 194–95; counseling, 94
Masculinity: aggressive, 64, 88; increased, in women with age, 149–
50; univocal, 94
Mastery Orientation: Active, 50–51, 55,
59, 62–63, 65–66, 84, 89, 92, 103, 107,
131, 133–34, 145–46, 149, 205–6, 223, 227,
269, 271–76, 278, 280, 282–83, 285–87,
289–96; Active, Moralistic Matriarchs, 139–40, 304; Active, Qualified, 51–53, 54–55, 59–60, 62, 269–73,
275, 279–80, 282–83, 285–88, 290–91,
294–97; Active, Rebellious Daughters, 139, 140–41, 144, 304; Bimodal,
50–53, 55–57, 60, 62, 269–70, 273–74,
278, 280, 282–84, 290; Magical, 50–
51, 54–55, 59, 63, 65, 89, 103, 121, 143–47,
149–51, 205–7, 209, 226, 271, 273, 275,
279–80, 282, 285, 287, 290, 294, 297,
299, 309, 312; Magical, Paranoid, 57–
58; Passive, 50–51, 56, 63, 65–66, 84,

89, 92, 103, 108, 133, 146, 155, 203, 205–
7, 218–20, 223, 225–26, 272, 274–76,
286–89, 291–94, 297, 299, 312; Passive,
Constricted, 53–54, 57, 66, 270–73,
279–80, 282, 284–85, 290, 296; Passive,
Maternal Altruists, 141, 144, 304;
Passive, Passive Aggressors, 142–43,
304; Passive, Sensual, 53, 60–61, 270,
272–73, 279–80, 282, 284, 288, 290, 296;
see also Thematic Apperception
Test (TAT)
Masturbation, 310
Matriarch–oldest son alliance, 168–73
Matriarchy, 181–82, 312; kin-tending,
215, 232, 235, 249; later-life, 156;
mothers-in-law and, 310; older
woman as family head and, 162–64;
urbanization and, 244
Maturation: genotypic specieswide,
234; later-life, urbanization as in-
hibiting, 249
Maya, Highland, 28–33, 37, 39, 41; Active Productivity and, 100; gerontocracy and, 217; mother-seeking
and, 119–20; orality and, 118; Passive
Mastery and, 54; personality, ego
aspects of, 259–62; personality, surgent features of, 257–59; Rorschach
test and, 257; sources of discontent
for, 102; TAT testing and, 99, 262,
271, 273, 276–79, 282, 289, 291–92, 296,
298, 301
Maya, Lowland, 29, 31–34, 37; Bimodal
Mastery and, 53; mother-seeking
and, 119; Passive Mastery and, 54;
personality, ego aspects of, 266–67;
personality, surgent aspects of, 265–
66; sources of contentment for, 101;
Spanish as contact language with
during psychological interview, 42–
43; TAT testing and, 62, 122, 127–29
Mead, M., 91, 164, 175, 310
Mears, C., 311
Meditation, 91
Meerloo, J., 96
Men: definition of beauty and, 99;
dreams of, 14; exuberance increases
with age in, 67; fantasy styles of, 47–
83, 99, 173; fear of losing manhood

Menarche, 159, 305

Meneghan, E., 201

Menopause, 162; *see also* Postmeno-
pause

Menstruation, 158, 160–62

Mernissi, F., 165, 167

Metaphor: castration as, 227, 262;
death as, 66; emasculation as, 162;
harem as, 80; house as, 78–79; hunger as, 112–14; machinery as, 78; Passive Mastery as, for death, 66

Middleton, J., 175

Modernization, 314–15; economics of,
236–37; erosion of gerontocracy and,
236–37; mythic power sources and,
239–41; rapid, 252; loss of wisdom
and, 238–41

Monogamy, 169

Morsy, S. A., 165

Moses, 229

Mother: artistic images of, 125; "bad,"
72; as comforter, 118–19; "good," 72–
74; intrusive, 147; as oral object, 112;
reverent feelings for, on part of son,
171; separation from, 233

Mothers-in-law, 147, 165–66, 168, 170–
72, 175, 310

Mountain habitat, centenarians and,
4

Murder: Magical Mastery and, 54; serial, 252

Murdoch, G., 191

Murray, H. A., 24, 45, 49, 59

Mysticism, 18

Myth, 9, 43, 229, 232–33, 236, 239–41, 249

Nadel, S. F., 87–88, 169, 174

Narcissism, 131, 185, 194, 204, 245, 248;
casualties of elderly, 250–51; transformations of, 197–98; unmodified,
316

Nash, S., 211

Navajo: Active Mastery and, 51; Active Productivity and, 100; Bimodal
Mastery and, 52; early memories
and, 121; Eastern, 26–27; English as
contact language with during psychological interview, 42; medicine
men and, 30, 238–39, 255; old-age as-

sistance programs and, 315; orality
and, 104–11, 115–17, 119; parenthood
and, 205; personality, ego aspects of,
255–57; personality, surgent features
of, 254–55; peyote and, 314; pleasure
and, 126–27; Qualified Active Mastery and, 52; Rorschach test and, 128,
308; sources of contentment for, 101;
sources of discontent for, 102; syntonic orality and, 243, 299–300; TAT
testing and, 57–58, 62–70, 99, 127–29,
271, 273, 275–82, 287–88, 291–92, 295–
96, 298, 301; Western, 26–27

Neocortex, 7, 187

Neugarten, B., 145

Neural structures, sex-linked, 48

Neurohormonal stimulation, 48

New Guinea, tribal society in, 162, 164,
174, 177

Newton, N., 199

Nigeria, tribal society in, 87

Nixon, Richard, 216, 230–31

Norms: productivity, age staging of,
92–94; social check, 92

Nurturance: fantasied, 295; female, 62,
156; social, 88

Nutrition, 4; cultural differences in,
116; orality and, 126

O'Connel, L., 207

Oedipus complex, 40, 165, 172–73, 175

Offspring-to-parent ratio, 186

O'Keeffe, G., 151–52

Old age, *see* Aging; Elderhood; Men,
older; Women, older

Omni-pleasure principle, 130–32

Omnipotentiality, illusions of, 194,
197–98, 244, 251

Ontogeny, permanent evolutionary
design of aging recapitulated in individual lives and, 4–6

Orality, 103–18, 121, 125–26, 129–31, 205,
243–44, 254, 256, 258, 263, 299–300,
308–9

Orgasm, 96–97, 130, 212

Parenthood: adulthood and, 186; culture-tending and, 231–34; emeritus,
by elders, 214–34; exit from active,

Species: freedom of, 186–87; human, as distinct entity, 10
Spencer, P., 84, 168
Spitz, R. A., 311
Steed, G., 89
Steig, W., 72–74, 77–78, 125, 309
Streib, G., 307
Strong Test, 128
Suicide, 54, 252
Superego, 197
Supernatural, 28, 85–86, 169, 175, 220–21, 277
Syntonic orality score (SOS), 104–12, 115, 206, 243, 299–300

Taboo, 27, 89, 219–22, 237, 313; food, 161, 308, 313–14; incest, 3, 216, 256; language, 161; menstrual, 162
Tachibana, K., 149–50
Tai Fu Ku, 98
Talmon-Garber, Y., 93
Tatum, R., 181
Tectonics, plate, 11–12, 20–21
Testosterone, 48, 182
Thailand, peasant society in, 90
Thanatos, 185; female, 156, 174; male, 98–99, 115
Thematic Apperception Test (TAT), 24, 44, 46, 83, 99, 101, 103–4, 148, 222–23, 225, 243, 306, 308; Bats and Man card, 122–24, 262, 293, 295–98; Boy on Cliff card, 126–28, 301; denial and, 121; Desert Scene card, 122–23, 267, 293–97; Druze version of, 45; Family Scene card, 122, 135–38; Farm/Family card, 142; 208, 302–3; Heterosexual Conflict card, 49, 59–62, 64–67, 72, 143, 150, 199, 286; Horse and Men card, 66, 150, 267–69, 272–75, 286, 292; Indian version of, 45; longitudinal profile, 57–59, 63–71, 110, 112, 137, 272, 274, 280–81, 289–91; Male Authority card, 66, 135–36, 139, 275, 277, 282–83, 285–92; mother-seeking and, 119–20; Rope Climber card, 45, 49–53, 54–57, 59, 63–64, 126, 143, 150, 271–72, 275–81, 286, 292; see also Mastery Orientation
Thomae, H., 201, 307–8

Toilet training, 189, 265–66
Thurnher, M., 156
Tobin, S., 145
Total orality score (TOS), 104, 108, 116, 308
Totem and Taboo (Freud), 40
Townsend, P., 309, 315
Triaxial view of development, 10–12, 16, 19–21
Tribal society: Ainu, 176; Amazonian, 308; Aranda, 176; Arunta, 313; Asmat, 162, 164, 166, 174; Attawapiskat Cree, 307; Australian bush, 313; Bemba, 168; Blackfoot, 160; Chippewa, 113, 162–63; Chukchi, 113, 219; Comanche, 89–91, 176, 306; Eskimo, 160, 308, 313; Galla, 86; Groot-Eylandt, 88; Gusii, 169; Hare Indians, 313; Ibo, 87, 314; Ik, 315–16; Iroquois, 113, 310; Kikuyu, 175, 308; Kipsigis, 86; Korongo, 87–88; Kowranega, 313; Kwakiutl, 113; Lele, 85; LoDagaa, 158; Maori, 178; Masai, 84; Maya, 28, 31–34, 37, 39, 41; Mayotte, 179–80; Mesakin, 87–88; Navajo, 113, 115–17, 119, 121, 126–29, 217, 238–39, 243, 254–56, 308, 314–15; Ngecha, 169–70; North Piegan, 160; Nupe, 169, 174; Nyakusa, 85; Omaha, 113; Pomo, 95; Rendille, 85; Salish, coastal, 160; Samburu, 84–86, 89, 167–68, 306; Samoan, 161; Sherdukpen, 177; Sioux, 313; Tallensi, 176, 310; Tonga, 162, 170; Yanomamo, 160–61; Yukaghir, 113; Yusufzai Pakhtun, 166; Zanaki, 113; Zulu, 85, 176; see also Society
Turnbull, C., 315–16

Urbanization, 236–37, 242–44, 249, 314–15
Urination, 161, 221

Van Arsdale, P., 162–63, 166, 174
Vatuk, S., 211
Vegetarianism, 91

Wang Chi, 184
Weaning, 189
Webber, I., 93

Index